# PRAISE FOR
# *NEUROSCIENCE FC*
# *ORGANIZATIONAL*
# SECOND EDITION

'This book is a great synthesis of scientific research and practical implementation. It enables readers to understand the science and then apply it to improve both their own performance and that of their teams.'
**Vincent Walsh, Professor of Human Brain Research,
University College London**

'This is an outstanding read, packed full of compelling evidence and presented in short, digestible chunks with practical suggestions and ideas. It's an invaluable handbook for any HR professional looking to get under the human skin of organizational change and to help them to make sense of the stuff we know intrinsically to be true.'
**Dr Neil Wooding, Chief People Officer, Ministry of Justice**

'This book is a must-have resource for change practitioners and leaders who want to understand how people process information, what can be done to manage threat and reward responses, and how to lead change. The way we manage change is changing and this book helps us understand that every person travels a unique and personal journey through change that leaders can influence when the right approach is used.'
**Michelle Pallas, USA Country Co-lead, Change Management Institute**

'In this second edition, Hilary has added significant insights on behavioural change (both for self and others) as well as clear applications readers can use to maximize their own productivity. This book continues to provide leaders and change agents with a timeless version of how to translate science into practical steps in working with change and leading their people.'
**Dr Mee-Yan Cheung-Judge, Visiting Fellow, Roffey Park Institute and the Civil Service College Singapore; recipient of two Lifetime Achievement awards; voted top influential thinker in 2018 by *HR* magazine**

'Hilary Scarlett has achieved something remarkable – an immensely readable guide to the way that modern work impacts upon our minds. Anyone wanting to improve the impact of their job on them – or do the same for their team – will be enthralled by it.'
**Bruce Daisley, VP EMEA, Twitter; author of *Sunday Times* bestseller *The Joy of Work***

'Hilary Scarlett has produced a fascinating and accessible guide to the science behind our behaviour at work. Full of practical advice, this will be a useful book for anyone wanting to become a better manager, and for leaders with the vision to build a more positive workplace.'
**Jo Swinson CBE, Minister for Employment Relations 2012–2015; Deputy Leader, Liberal Democrats**

'The additional work in the second edition of *Neuroscience for Organizational Change* adds an extra dimension of practicality to the ideas in the first edition. Put simply, the extra chapters gave me lots of ideas for the content of my change plans, to really help achieve behavioural change. I loved the emphasis on well-being and how to treat our brains with care. I think this guidance is a great conversation starter for organizations beginning to prioritize mental health at work. This book is useful, relevant and readable.'
**Melanie Franklin, Director, Agile Change Management**

'Filled with pertinent research and generous insight, this book is a pleasure to read. Every professional involved with behaviour change should read it to ensure that they have the latest applied information on the topic. Hilary Scarlett has seamlessly integrated the logic for why it is important to change with the realities of how our brains can fight us in changing. Working with behaviour change without the scientific knowledge provided in this book is a certain handicap for all kinds of change professionals.'
**Theresa Moulton, Editor-in-Chief, *Change Management Review***

'*Neuroscience for Organizational Change* is a key resource for managers and consultants in the planning and conduct of neuroscience-based organizational change. Hilary Scarlett provides a comprehensive and practical discussion of the link between neuroscience principles and organizational change. This is a book that you will use again and again.'
**Walter McFarland, Board Chair Emeritus, Association for Talent Development (ATD); co-author of *Choosing Change***

'Change, in all forms, wants to disrupt and do things differently, while our brains prefer a steady state with repetitive actions. Hilary Scarlett has provided a practical, accessible way to understand this reality and improve how we manage our leadership and change needs.'
**David W Jamieson, Professor and Director of Organization Development and Change Doctorate, University of St Thomas, Minneapolis**

'Change is the only constant in business (and life), so the more we can each understand how the human brain deals with change from a neuroscience perspective, the better we're able to create the conditions for employees to thrive during this uncertainty. Understanding the science behind why we react the way we do to change is fascinating, plus it will significantly increase the chance of your change projects landing successfully. I have recommended this book to many friends, colleagues and peers. Read it!'
**Kim Wylie, Global Director of People Development, Farfetch; former leader of Global Change Management Practice, Google Cloud**

'This book is the embodiment of the potential of organizational neuroscience. It takes the reader on a journey of discovery into the inner workings and mysteries of how our brains work and influence behaviour in a way that is deep and technically accurate, yet digestible by anyone. Each chapter provides a stand-alone road map for understanding and confronting specific brain phenomena that impact different aspects of organizational change, and the new chapters in this second edition contain further useful real-world applications.'
**William Becker, Associate Professor of Management, Pamplin College of Business, Virginia Tech**

'A superb overview of neuroscience and its application to leadership, communication and organizational change. Hilary Scarlett elegantly plots a path through the growing body of applied research and makes it accessible through real-life examples, workshop models and questions for reflection. This second edition adds a useful chapter on individual productivity and mental well-being. With CEOs and boards increasingly seeing organizational performance as dependent on the emotional engagement and mental health of their workforce, this book is a great source of guidance and inspiration.'
**Phil Askham, Global Head of Employee Communications, HSBC**

'This book is a must-read and a must-use! For those of us practising the art of communication and change with people at work, Hilary Scarlett brings insights from solid neuroscience to illuminate and prompt new thinking – with realistic and practical applications, for ourselves and for those whom we aim to guide and support.'
**Lindsay Eynon, Partner and Change Communications Manager,**
**John Lewis & Partners**

'A brilliant read for anyone leading change. This book really helps you understand how people tick and is packed full of practical advice to guide people through uncertainty and have change land successfully.'
**Mairi Doyle, Director of Internal Communications, Bupa**

'Hilary Scarlett's *Neuroscience for Organizational Change* is a delight to read: eminently accessible and useful to both managers and students interested in change. The book offers a compelling blend of theory with practical application and gives the reader a perfect road map forward.'
**Paula Payton, Lecturer in Applied Analytics, School of Professional Studies, Columbia University**

'Hilary Scarlett speaks directly to a lay audience to intelligibly, openly and powerfully outline what neuroscience means and how important understanding it is. Her book is an engrossing read for anyone attempting to initiate change in an organization, or, indeed, for all of us working with people in a changing world of work. It is now my go-to guide for welcoming even the most experienced executives into neuroscience's application in human resources.'
**Jessica Hayes, Head of People and Talent, London**

**Second Edition**

# Neuroscience for Organizational Change

An evidence-based practical guide to managing change

Hilary Scarlett

KoganPage

First published in Great Britain and the United States in 2016 by Kogan Page Limited
Second edition 2019

| | | |
|---|---|---|
| 2nd Floor, 45 Gee Street | 122 W 27th St, 10th Floor | 4737/23 Ansari Road |
| London | New York, NY 10001 | Daryaganj |
| EC1V 3RS | USA | New Delhi 110002 |
| United Kingdom | | India |
| www.koganpage.com | | |

© Hilary Scarlett, 2016, 2019

Hardback     978 1 78966 005 0
Paperback   978 0 7494 9318 9
Ebook          978 0 7494 9319 6

**British Library Cataloguing-in-Publication Data**

A CIP record for this book is available from the British Library.

**Library of Congress Cataloging-in-Publication Data**

Names: Scarlett, Hilary, author.
Title: Neuroscience for organizational change : an evidence-based practical
   guide to managing change / Hilary Scarlett.
Description: Second Edition. | New York : Kogan Page Ltd, [2019] | Revised
   edition of the author's Neuroscience for organizational change, 2016. |
   Includes bibliographical references and index.
Identifiers: LCCN 2019012513 (print) | LCCN 2019014178 (ebook) | ISBN
   9780749493196 (Ebook) | ISBN 9781789660050 (hardback) | ISBN 9780749493189
   (pbk.)
Subjects: LCSH: Organizational change–Psychological aspects. |
   Organizational behavior. | Cognitive neuroscience.
Classification: LCC HD58.8 (ebook) | LCC HD58.8.S2935 2019 (print) | DDC
   658.4/06–dc23
LC record available at https://lccn.loc.gov/2019012513

Typeset by Hong Kong FIVE Workshop
Print production managed by Jellyfish
Printed and bound by CPI Group (UK) Ltd, Croydon, CR0 4YY

# CONTENTS

# PREFACE

I have worked in the field of organizational change for many years, and in 2009 came across an article which touched on the insights neuroscience could bring to organizations. I found it fascinating, inspiring and practical. At long last, here was science and evidence being brought to change management. Neuroscience brings a new lens through which to look at and understand people. Since then I have been on a bit of a mission: I passionately believe that if organizations, leaders, all of us, understand a little bit more about how our brains work then we can work with the brain, not despite it. There is a real 'win–win' here: when organizations and leaders understand better what our brains need to focus and perform at their best, this will help productivity. Better still, in a time when there is much concern about mental and emotional welfare, we can reduce stress by designing organizations and work in a more brain-friendly way. We need to change the way we introduce and manage change in organizations. Neuroscience will enable us to do this.

This is the second edition of *Neuroscience for Organizational Change*. Although the field of neuroscience is continuing to grow and there are now many more studies than when I wrote the first edition, what we know about the brain and what I covered in the first edition have not fundamentally changed. I have therefore updated parts of Chapters 1–9, but they have not radically changed. There has been what is called the 'replication crisis' in psychology, where some long-standing pieces of research are being debated. Where relevant, I have reflected this in the text. I have also added in new research.

I have added two new chapters to this edition. Chapter 10 looks at how we help people to change behaviour. Chapter 11 explores how we can help ourselves and others plan the working day to get the best out of our brains. Since writing the first edition I have had the pleasure of working with many people in different organizations. In Chapter 12, three such organizations share how they have been using and applying neuroscience.

# ACKNOWLEDGEMENTS

I'd like to thank all the neuroscientists, psychologists and authors whose research I have referred to in this book. Without their work, this book would not exist and our approach to managing change would still be largely based on hunches, opinions and past experience – which are not necessarily bad, but much better to counterbalance them with evidence.

Once again, I'd particularly like to thank Dr Erman Misirlisoy, formerly of University College London (UCL) and now based in the USA. This book is all the richer thanks to our conversations and Erman's suggestions of research studies to include. Any errors in this book are mine, not his. Thank you to Professor Vince Walsh, Head of Brain Research at UCL, who reviewed the content of the original masterclass and was generous in adding content. Thank you to Dr Pranjal Mehta, also of UCL, for his insights on stress, hormones and their impact on decision-making and behaviour, and also for sharing the 'Best in Show' research (see Chapter 6), which is not only insightful but also makes me smile whenever I think about it. Thanks to Dr Björn Christianson at UCL for making the introductions.

Thank you to all my clients. Their feedback continues to reinforce my belief that bringing insights from neuroscience to organizations makes a significant and beneficial difference, and that it helps not just at an organizational level but at a personal level too.

Thank you, Michael Pounsford. Mike has not only been a great colleague for many years but a good friend too. Mike and I worked together on the research referred to in Chapter 12. 'Thank you' also needs to go to Jennifer Cramb – it was a conversation with Jennifer that sparked the thought of writing the first edition.

Thank you too to all those people on social media who have endorsed the book and shared what they have gained from it. Thank you for taking the time to let me know. I really appreciate it and your comments have spurred me on while writing this second edition.

Thank you to James Bromet and Charles Cavalla for your help with the words – editing, proofreading and checking the references. Particular thanks to James, who has had to witness the pains and pleasures of my writing this book.

I am grateful to the team at Kogan Page for their enthusiasm, charm and perseverance in asking me to write this second edition.

Last but not least, thank you for buying or reading this book. I am always interested to hear how people are applying the learning and what difference it has made, so please let me know via our website: www.scarlettandgrey. com

*For James and Olivia*
*and in memory of Geoffrey*

# PART ONE
# The challenge

# Introduction to neuroscience 01

**W**ork has changed hugely but our brains have not, and therein lies the challenge. Organizations need their people constantly to adapt and change, collaborate, innovate and perform at their best. But our brains are not designed for 21st century organizational life: they think they are still out on the savannah, and for our brains the goal is all about survival. Given these constraints, those leading organizational change need to understand how the brain perceives and processes change, and what can be done to enable our brains to work at their best during times of uncertainty. This book provides insight on how to do that.

## Why should organizations be interested in neuroscience?

Perhaps the question should be put the other way around: why wouldn't organizations be interested? Change is difficult. We know from experience that personal change isn't easy, from taking more exercise, eating more healthily, getting enough sleep, starting a new role, moving to a new location, to learning a new skill. It doesn't come easily to us. Organizational change is even harder: trying to persuade tens, hundreds, thousands and tens of thousands of people to change is a challenge. There probably isn't a single organization that has not been through change, be it the impact of globalization, restructuring, new technology, changing legislation or shifting customer expectations. Moreover, as many employees identify, and as we know from personal experience, it's not just the big change programmes that are difficult – it is the constant layer upon layer of small changes that gradually wear us down.

Despite this vast experience of change, industry and employee surveys tell us organizations could do better. Time and again industry reports reveal how many change programmes fail to deliver the promised efficiencies and benefits. The focus on employee well-being is in part a recognition that it is in no one's interest in the long term to keep employees in a constant state of stress. It is clear that we can, and need to, get cleverer at helping organizations and people through change. Neuroscience can help us do this.

Neuroscience, the study of the nervous system including the brain, provides a new lens through which to look at change. There are different theories of change but there is something very appealing and practical about understanding how our brains deal with change and what would help them to become, and remain, resilient. Neuroscience helps us to understand ourselves and others better. For some leaders I work with, just having this new perspective on change and deeper understanding of people has been enough to make a significant difference to their ability and confidence to lead others through change. Understanding more about our brains, what drives them and what gets in the way, has helped both to guide their actions and build their empathy for others. Some of them will tell their stories later in this book.

So, neuroscience, as we'll explore in Chapter 3, helps to explain why we find organizational change difficult. More interestingly and more importantly, it provides clear guidance on what practical things we can do to help people through change (and that is the focus in the chapters in Part Two of this book). Although still in its infancy, it is already proving immensely useful in bringing to light what enables us to be focused, to learn and to perform at our best.

Part of the appeal of neuroscience is that it resonates with people's experience: it explains why we find uncertainty disturbing and distracting. It helps us realize that our reactions and emotions are 'normal' and people find this reassuring: there's nothing wrong with them, it's just the way our brains deal with change. It also underpins the intuitions of good organizational change practitioners and leaders: for example, that having a good relationship with a line manager is not just 'nice to have' but does make a real difference to us and enables us to think and work better. Neuroscience shows that this is not just a matter of opinion: there are biological reasons as to why we feel this way.

Another benefit is that although at one level neuroscience is complex, at another level it is relatively easy to apply. You don't need to be a neuroscientist. Nor do you have to wait for major culture change in your organization. Every leader, manager and employee can get on and apply it straight away. Nor does it necessarily require radical change: small actions can make a significant difference to our ability to focus at work. The learning can be applied at a local level by each leader with their team, or at a macro level, right across the business. By understanding our brains and the way they function, we can work with the physiology, not fight it.

One more reason why organizations are finding neuroscience so useful is that it provides scientific evidence and a language which appeals to even the most sceptical of leaders. People I work with have said how useful it is when working with their leadership teams to be able to draw on science, to refer to research and evidence. This was brought home to me when I was working with a group of leaders in a bank who had some demanding targets to meet while employees were facing redundancy. During one session a banker commented 'I like this. It's not the usual psychofluff I get from HR and Communications. This is *science*.'

## Why so much interest right now?

There's no doubt that neuroscience is coming to the fore. In both the United States and Europe, governments have committed large budgets so that we can better understand the brain. In the United States, The Human Connectome project is mapping the brain's neural connections. You can find out more at **http://www.humanconnectomeproject.org/**. Another example is The Wellcome Trust's and the Education Endowment Foundation's *Education and Neuroscience Initiative* in the UK which is evaluating the impact of education interventions that have been informed by neuroscience. One of the reasons there is so much interest right now is that we are just beginning to see neuroscience come out of the lab and into the workplace: we are just beginning to explore the practical applications.

For the majority of organizations, people's brain power is crucial. Many organizations and indeed many countries have moved away from agriculture and manufacturing. Their survival and success depend on people's ability to think, innovate, adapt and collaborate. Brain power is all they have.

Change is speeding up – we only need to look back at the last 20 years or so to see how much change there has been and how quickly new changes have arrived. Facsimile machines have come and more or less gone; videos have almost disappeared. The devices and means we have to communicate and share information would have amazed us 30 years ago. Successful organizations need to be able to adapt swiftly.

In short, here are 10 reasons why neuroscience matters to organizations:

1 We need to change the way we manage change in organizations: by understanding the brain, we can do this better.

2  Neuroscience brings a new lens through which to look at and understand people and what motivates them.

3  At long last, neuroscience brings scientific evidence to leading people, change management and employee engagement, proving they are not 'soft'.

4  It resonates and underpins the instincts of successful leaders; it builds confidence as to what to do to support people through change.

5  Performance: neuroscience is all about what helps our brains to perform at their best.

6  It's practical and can be relatively easy to implement.

7  When we understand how our brains work, we can work with the physiology, not fight it.

8  It's useful at the macro and micro level: from how to plan change in an organization to how to organize your day to allow your brain to think at its best.

9  There is a greater focus on well-being in organizations, physically, mentally and emotionally: organizations owe it to their employees to understand what will help keep their brains and minds in a better state.

10  It provides science and evidence and these can be very persuasive to sceptical leaders.

# Key moments in the history of neuroscience

Neuroscience is very much in the news in the 21st century, but it's not new. For thousands of years, people have wondered what the purpose of the brain is. Here are a few of the key milestones and characters who have contributed to our learning.

## The pioneers of neuroscience

The ancient Greeks were interested in the role of the brain but held widely differing views. Some believed that the brain was the seat of the soul but that it had nothing to do with thinking; they believed that thought was in the heart. The logic was that the heart was seen as the source of blood, and if you lost blood, that was the end of life and of thinking. Aristotle thought the brain's task was simply to cool the blood. The Egyptians seem to have placed little value on the brain: although they preserved the body for the afterlife, they scooped the brain out of the head and discarded it.

Several ancient Greeks made a significant contribution to our knowledge. In the 5th century BC, Alcmaeon of Croton found a connection between the eyes and brain (probably the optic nerve) and so began to establish a link between the brain and mental activity. In around 400 BC, the physician Hippocrates of Kos recognized that epilepsy and madness were disturbances of the brain. Hippocrates also identified that a cut on one side of the brain caused a spasm on the other side of the body. This is what we now recognize as lateralization of brain function: that the left side of the brain controls the right side of the body and vice versa. Around the 3rd century BC, Erasistratus identified what we now know as motor and sensory nerves from the body to the brain. The Roman physician and anatomist Galen (129–199 AD), an important figure in ancient science, noted the impact of a brain injury on mental activity.

Like the Greeks, it seems that the ancient Chinese and Indians also placed greater significance on the heart as the organ ruling over feelings and sensation. Texts exist that show that the Indians noted mental disorders as a consequence of head injuries. The Chinese suggested that different emotions were spread across the internal organs, with the spleen being important to consciousness, the liver to anger, the kidneys to fear and the heart to happiness (Finger, 1994).

The views of the Greeks are important as they influenced so much of what people in Europe thought until the Renaissance. Understanding was moved along by Leonardo da Vinci whose work included the dissection of the brain of an ox. In the 17th century, René Descartes played a role too. Descartes distinguished between the mind and the body: he made explicit that the bodily functions of the brain are completely separate from the mystical qualities of the soul: this freed people to think about the practical functioning of the brain without fearing what religious authorities might say.

## *Phineas Gage*

In September 1848 Phineas Gage was working near Cavendish in Vermont. He was the well-liked and capable 25-year-old supervisor of a group of men working on the new rail line. They were blasting through rock when Phineas decided that he would take care of pouring gunpowder into a deep and narrow hole that they had drilled into the stone. He rammed a long iron rod to tamp down the gunpowder but a spark was accidentally ignited and the tamping iron, which was over a metre long and over two centimetres wide, blasted upwards. It hit Phineas just below his left eye and went through the top of his head and landed some 50 metres away. Amazingly, Phineas survived and he made a good physical recovery. However, his personality completely changed. From being a good worker and an even-tempered man, he became unreliable and unpredictable. He could no longer work at his job and his marriage broke down. John Harlow, one of the doctors caring for him, wrote, 'He is fitful, irreverent, indulging at times in the grossest profanity (which was not previously his custom) manifesting but little deference for his fellows, impatient of restraint or advice when it conflicts with his desires, at times pertinaciously obstinate, yet capricious and vacillating, devising many plans of future operation which are no sooner arranged than they are abandoned... he was "no longer Gage".'

The incident was disastrous for Phineas, but his accident demonstrated that certain parts of the brain influence both our behaviour and our very personality.

## Paul Broca and Karl Wernicke

Just a decade later in 1861, Paul Broca, a French neuroanatomist and anthropologist, examined the brain of a dead man. The man's name was Leborgne but he had been nicknamed 'Tan', as over two decades his ability to speak had reduced to the stage where all he could say was 'Tan', although his other mental abilities were intact. Broca's examination showed that Tan had a lesion on the left side of his frontal lobe, and Broca deduced that this had had an impact on Tan's reduced ability to speak. This area is now known as Broca's area and we will refer to this again in Chapter 8 on communication and storytelling. In 1876 Karl Wernicke, a German physician and psychiatrist, noted that not all communication problems were a result of damage to Broca's area. He found another area on the left of the brain that, when damaged, limited people's ability to understand language. This part of the brain is now known as Wernicke's area.

As a side note, in a study by Pujol *et al* (1999) 96 per cent of right-handed people were shown to have language in the left hemisphere, whereas only 76 per cent of left-handers have communication on the left; 14 per cent of left-handed people have bilateral activation and 10 per cent showed dominance in the right hemisphere.

## World War One (WW1)

World War One (WW1) also helped to develop our knowledge. Much of what we have learned about the brain stems from people's misfortune – when things have gone wrong through natural causes or because parts of it have been damaged or destroyed. WW1 marks a point when medicine had become sophisticated enough to prevent some soldiers dying of head wounds, enabling them to live but with mental, emotional, behavioural or personality disorders.

## 'HM'

Just as Phineas Gage never expected to feature in the history of neuroscience, nor did Henry Molaison, better known as 'HM'. HM was born in 1926 and he developed a very severe form of epilepsy – so serious that he could not work or lead a normal life. So in 1953, when he was 27 years old, HM underwent an operation to relieve the epilepsy. This involved removing certain parts of the brain, including the hippocampus. This operation had been conducted before, but on previous patients only one side of the brain was operated on. HM's epilepsy was so severe that the surgeon decided to remove parts on both sides of his brain. The result was devastating; so much so, that this procedure has never been repeated. For HM, after the operation, was incapable of forming new 'explicit' memories: he was trapped in the present. This meant that he could not remember people he had met since the operation. One of the researchers working with him related how she had to reintroduce herself each day and, even if she left the room just for a short while, when she returned he would have forgotten that they had ever met. He was able to retain memories for about 30 seconds and then they were lost.

As for memories formed before he had the operation, Henry retained some but not others. He had lived through World War Two (WWII) and could remember facts about it, but he began to lose autobiographical memories such as his experience at high school. We might think of past memories as being similar, but HM showed that the brain processes the recall of semantic memories, eg facts, in a different way from the recall of episodic memories such as unique events from our personal lives. From HM we have learned that the retrieval of autobiographical memories requires the hippocampus.

HM did retain the ability to form some new memories: 'motor tasks', ie any task that involves learning new skilled movements (as opposed to facts).

For example, HM was asked to learn how to trace between two outlines of a star: he improved considerably each time he was asked to perform the task, although he could not recall ever having done the task before.

HM unwittingly taught us the role of parts of the brain, in particular the hippocampus in creating and retrieving new memories.

## Technology

In more recent years our knowledge has advanced again thanks to technology. X-rays have been around since the early 20th century. Newer machines, such as Functional Magnetic Resonance Imaging (fMRI) scanners, do not use X-rays and so are considered safer. They enable scientists to look at the brain as we are involved in tasks. fMRI measures changes in blood oxygen concentration; it uses the fact that oxygen is carried by haemoglobin, and the amount of oxygen carried affects the magnetic properties of haemoglobin. fMRI has helped neuroscientists to identify that a range of brain areas might be activated by a specific task. fMRI shows that if the nature of a task is changed slightly, eg reading words rather than hearing words, different parts of the brain are activated.

Technology is moving on again as it becomes more mobile: tools such as electroencephalography (EEG) caps measure electrical activity on the surface of the brain and enable research to take place outside the lab and while people are performing tasks in the workplace.

# Caveat

## Misinterpretation

There is a lot of hype, over-interpretation and misinformation out there; 'neurononsense' is one of the kinder ways of describing it. One example of this is an Op-Ed article in the *New York Times* in 2011 called 'You love your iPhone. Literally.' The researcher, who worked for a branding agency, put 16 participants into an fMRI scanner and looked at which parts of the brain were activated when they saw or heard an iPhone. Amongst other things, the researcher claimed that, because the insular cortex was activated, 'The subjects' brains responded to the sound of their phones as they would respond to the presence or proximity of a girlfriend, boyfriend or family member'.

Really? Many academics were quick to respond. The insula does respond to emotions. In fact as many as one-third of neuroimaging studies are said to show activation in the insula. It can be triggered by positive emotions but the primary function attributed to the insula is disgust, not love. It responds to disgusting tastes and images, or when we see something we think is morally wrong. It responds to pain. This is one of many examples of misinterpretation: it is dangerous to assume that because a brain region can be activated by, for example, love, that activation there always means that the participant is feeling love.

This is by no means an isolated case of over-claiming for neuroscience. The areas of marketing and advertising seem to be particularly prone to overexcitement. In an advertising film for a fast car, the manufacturers stated that scans showed that it is almost as exhilarating to drive their car as to fly a fighter jet. They said that they had proved this by asking the driver to wear an electroencephalography (EEG) cap and showed that the driver's levels of a chemical called dopamine (a neurotransmitter linked to positive experiences that we will refer to later) had risen to almost those of someone flying the jet. There's a major flaw in this argument: EEG caps measure electrical activity in the brain and can't measure dopamine.

There is a great deal of enthusiasm – understandably so – for trying to identify what happens in people's brains before they decide to buy a product. If only we could put people into a brain scanner and see which messages will convince them to buy, what a difference that would make. The well-known saying, 'Half the money I spend on advertising is wasted; the trouble is I don't know which half' would no longer be true. But we are not there yet.

## It's early days

Despite identification of parts of the brain going back thousands of years, neuroscience is still in its infancy. There is still a huge amount we do not know, and possibly some things we will never know, such as 'the hard problem' of neuroscience: how does a lump of tofu-like substance inside our heads give rise to the mind and to consciousness?

The 2014 Nobel prize-winning work of Professor John O'Keefe, Director of the Sainsbury Wellcome Centre for Neural Circuits and Behaviour and Professor of Cognitive Neuroscience in the Department of Cell and Developmental Biology, Division of Biosciences at University College London (UCL), changed our understanding again. The hippocampus is a part of the brain very much associated with memory (as HM's story illustrates) but Professor O'Keefe's work showed that the hippocampus is not just about memory but also about spatial orientation. His work together with that of May-Britt Moser and Edvard Moser demonstrated that the hippocampus has a role as the brain's positioning system, an internal GPS so to speak. Each of these discoveries shifts our understanding and causes us to think again about assumptions we have made about the brain.

## *The lab is not the workplace*

Much neuroscience research is still done in a lab and, to point out the obvious, the lab is not like the world of work. In an article in 2015 (**http:// ideas.ted.com/how-scientists-make-people-laugh-to-study-humor/**), Sophie Scott, Deputy Director of UCL's Institute of Cognitive Neuroscience, discusses the challenges she faces in researching communication and, in particular, laughter: laughter is not an easy thing to create in a scanning machine. So, we also need to be aware that much research is done in a lab which is a very specific and peculiar environment. In addition, often research participants are inside large machines, such as fMRI scanners, which are very far from the typical workplace. Moreover, there might be biases because of where the research is done (the richer countries in the world) and the participants (often university students). However, this is changing as technology develops and becomes smaller, more mobile and more portable.

## *This book*

What we know and understand about the brain is constantly changing. That said, there are areas of knowledge that are now well established and that the great majority of neuroscientists would agree on. In this book, I have, by and large, kept to well-researched and well-established areas.

# About this book – how will it help?

This book is all about helping those responsible for leading change to have a better chance of doing so successfully. Change is speeding up and there is more and more pressure on leaders and managers to do it quickly and to do it right first time.

In part, this book is about sharing understanding and providing a new lens through which to look at people and change. For some people just understanding more about the brain and how it perceives the world is useful and is enough. But applied neuroscience is also very practical and small actions can make a big difference. The book is a balance of science, practical examples, personal stories and questions to help the reader plan change from a brain-friendly perspective. These first three chapters provide context and a foundation for the rest of the book. The chapters in the second part each focus on an aspect of organizational change, share some of the science that helps us to understand what is going on, and set out some very practical things we can do to help. It's a book that you can read from cover to cover and/or dip in and out of, as you choose.

# Summary of key points from this chapter

- Work has changed significantly but our brains have not: they are not designed for the 21st century workplace.
- Organizational change is ubiquitous but we are still struggling to do it well.
- Understanding our brains and how they deal with change means that we can work with the physiology, not fight it.
- Neuroscience provides a lens through which we can better understand ourselves and others.
- It provides science and evidence which can be very persuasive.
- Although at one level it is complicated, it can be translated into practical actions that make a difference.
- Neuroscience is still in its infancy, but an interest in our brains is not new.
- Until the arrival of tools such as fMRI scanners and EEG caps much of our knowledge stemmed from the damaged brain.
- Neuroscience is the subject of much hype, especially in the areas of marketing and advertising, and we do need to tread with caution.

# References and further reading

http://www.humanconnectomeproject.org/

Corkin, S (2014) *Permanent Present Tense*, Penguin, London

Finger, S (1994) *Origins of Neuroscience: A History of Explorations into Brain Functions*, Oxford University Press, Oxford

Pujol, J *et al* (1999) Cerebral lateralization of language in normal left-handed people studied by functional MRI, *Neurology*, 52

Scott, S (2015) http://ideas.ted.com/how-scientists-make-people-laugh-to-study-humor/

# Brain facts 02

If we understand how the brain works, then we can begin to change it. This chapter takes a look at the structure of the brain and some key principles that govern how it works. Understanding these principles makes it easier to understand what drives our behaviour and what helps us to focus at work.

## Our brains

The brain weighs about 1.3 kg and has the consistency of tofu or a soft-boiled egg. It is very compact and its surface area is made much larger by having ridges and dips called gyri and sulci. The larger surface area allows us to do more complex tasks and thinking (the brains of cats, for example, are much smoother by comparison). The brain has no moving parts but brain cells do migrate as the structure of the brain develops. The brain has no pain receptors but it has a 'brain map'. This means that when you hurt a particular area of the body, that part of the map in the brain becomes active, and this leads to the localized experience of pain in the body part itself. The brain also has a 'motor map' and a 'somatosensory map' where different brain areas correspond to different body parts.

We don't have the largest brains on the planet – elephants have larger brains – but we do have brains about five to seven times larger than would be expected for a mammal of our body size. Our brains at birth are still only about 30 per cent of the size they will be in adulthood, but brains make up most of this difference in the first year. By the time we are three, our brains are about 80 per cent of adult size and this increases to about 90 per cent by the age of five.

### Why have a brain?

Why do we have a brain? What's its purpose? Lots of living beings don't have brains – plants, for example. Some scientists argue that the purpose of the brain is movement: brains give us the ability to navigate the world. A creature called the sea squirt is often given as an example of this: while it is moving around in the sea, it has a simple brain, but once the sea squirt finds a rock to attach itself to, its brain disappears. Why keep a brain if you no longer need it? We now know that the brain's functions have extended far beyond movement. It is responsible for our psychological selves and personalities.

## Why have our brains become larger?

Up to about two million years ago the relative brain size of our ancestors was the same as that of the great apes today. However, something happened in evolution to cause our brains to grow to three to four times the size of those of our ancestral apes. Why evolve a larger brain? So that we could move around more, find more nutritional food such as berries and nuts rather than just leaves? So that we could better collaborate and hunt together? Or was it also a social need, not just a physical one? Robin Dunbar, evolutionary psychologist at Oxford University, argues that we evolved larger brains to handle our increasingly complex and large social groups and the social skills that they require (2007). This idea is known as the social brain hypothesis. However, there are other animals that live in large groups – wildebeest, for example – that don't have large brains, so there is more to this: it is not just the number of relationships but the complexities and subtleties of relationships that require a larger brain. Joan Silk of UCLA (2007) suggests our brains grew because we have developed the ability to acknowledge relationships between others. You need to be able to 'read' others if you are going to live with others. You need 'theory of mind' skills: the ability to put yourself in others' shoes.

## Neurons

Our brain has many different types of cell, with different roles, and they fall into two broad groups: glial cells and neurons. Glial cells outnumber neurons and they provide structural and metabolic support for neurons. Neurons are key cells in the brain: they receive, process and transfer information. They communicate with one another by sending and receiving electrical signals in response to a stimulus. Our brains have around 86 billion neurons. Unlike glial cells, mature neurons do not generally tend to regenerate. The brain is not the only place where we have neurons: there are also about 100 million in the gut.

The 'classical neuron' has three parts:

1 A cell body or 'soma'.

2 Dendrites: these are like branches on a tree and they receive input from other neurons.

3 The axon: a cable of varying lengths, from microscopic to some that run right down to our feet. The axon sends signals on to other neurons and does this at very high speeds, ranging from 1 metre per second to over 100 metres per second.

**Figure 2.1**    Neurons and synapse

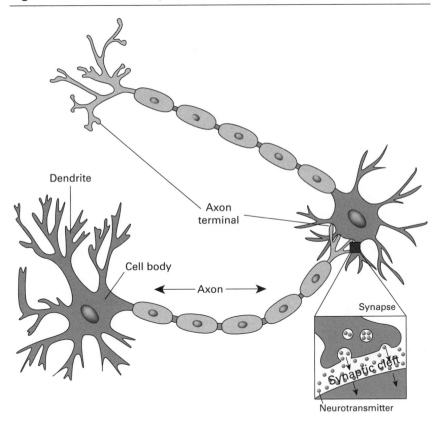

A foetus grows neurons at the rate of around 250,000 per minute and we are born with almost the full number of neurons we will have as adults, but the neural networks take more time to mature. The absolute number of neurons is not as important as the connections between them: the synaptic connections.

## *Synapses*

Many neurons do not actually touch each other but have a small gap between them called a synapse (sometimes also called the synaptic cleft). Neurons communicate by passing signals across this tiny gap and this signalling is done via chemicals called neurotransmitters. Connections are formed at the rate of 40,000 per second in babies and we form trillions of connections. Neurons that frequently signal to each other form a stronger connection – see more under *neuroplasticity*. As we will explore further on, it is not so much the number of neurons that matters but the number of synaptic connections between them. It's astonishing to imagine, but there are more potential connections (synapses) between neurons than there are atoms in the universe.

# Fundamental facts and principles about our brains

## *1 It's all about survival*

For our brains, the goal is survival, and they are very good at this; otherwise we wouldn't be here today. To survive, the brain needs to do two key things: avoid threats, such as the sabre-toothed tiger, and seek out rewards such as food and shelter. Both matter to our brains, but threat is far more important: we can go without shelter, food and even water for a while, but if the tiger gets us, that's the end of us. So, of the two, the threat response in our brains is far stronger than the reward response.

## *2 The impact of threat and reward on our brains*

Threat and reward responses have an impact on us both physically and mentally. Compared with the reward response, the threat response:

- kicks in faster;
- is much stronger;
- lasts for longer;
- increases heart rate;
- increases cortisol (stress hormone);

- reduces dopamine (a chemical that the brain finds rewarding – we'll come back to this later).

The threat response needs to be fast. For our ancestors it might have been a sabre-toothed tiger suddenly appearing; today it might be a car heading towards us at 50 mph. Either way, you need to respond very quickly to survive. When a loud noise surprises you, your heart rate increases automatically: your body is preparing itself for survival by preparing you to fight or to run away. Your blood is pumping round much faster to give muscles oxygen.

On the other hand, the reward response in our brains:

- is slower to be activated;
- is milder;
- is shorter-lived;
- increases dopamine.

The reward response is a nice feeling but it doesn't last as long. From a survival perspective it doesn't need to.

To get a sense of the two at work, think of a time when you received an e-mail criticizing you or your work. The emotion hits, physically and mentally, your heart rate increases, blood is pumped around your body faster and it is hard to think clearly: that's the threat response. On the other hand, think of the last e-mail you got praising you or your work. It's a pleasant feeling, but it doesn't have the same impact on you. It's nice to get the e-mail but your response is much milder than to that of the critical e-mail.

Although we might no longer be concerned about the sabre-toothed tiger, our brains are subconsciously constantly on the lookout for threats and rewards in the workplace. But especially threats. Organizations and leaders need to be aware of this extra sensitivity to threats. This is an important point to keep front of mind when we are going through organizational change: if the brain is much more sensitive to threat, then we need to think how we counterbalance this – more of this in Chapters 3 and 4.

## 3 Our brains are prediction machines

Our brains are helpless prediction machines. They are constantly, subconsciously, trying to guess what is going to happen to us. They want to be able to predict and make sense of the world. Again, this stems from the brain's drive to protect us. If the brain can predict what is about to occur, then it is in a better place to guard us and keep us out of harm's way.

To show how good your brain is at predicting and making sense, take a look at the following paragraph:

I cnlduo't bvleiee taht I culod aulaclty uesdtannrd waht I was rdnaieg. Unisg the icndeblire pweor of the hmuan mnid, aocdcrnig to rseecrah at Cmabrigde Uinervtisy, it dseno't mttaer in waht oderr the lteterts in a wrod are, the olny irpoamtnt tihng is taht the frsit and lsat ltteer be in the rhgit pclae.

Subconsciously, our brains are constantly predicting and interpreting. Take a look at Figure 2.2.

**Figure 2.2**   Convex and concave circles

The circles on the right look concave to us, and those on the left look con-vex. Now turn the book upside down. What happens? The concave circles now look convex and vice versa. This happens because your brain is inter-preting. The human brain, your brain, has learned that light typically comes from the sun and therefore from above, so due to the shadows, we see the circles on the right as concave.

Our brains have learned to predict because, on the whole, prediction is useful to us. Out on the savannah, it was useful to be able to predict that a rustle under some bushes might mean that there is a snake there. Prediction helps us to make decisions, and to make decisions much faster, subcon-sciously. It saves us from having to think long and hard about every issue we are confronted with. Why is this issue of prediction so important to those leading people through change? There are several reasons. The brain's desire to predict means that brains like information and certainty – if our brains have information, they are better able to predict. Every organizational change involves at least a little, and more often a great deal of, uncertainty and this prevents our brains from doing what they like to do – predict. More on this in Chapter 3.

Prediction – not having to consider carefully every decision – also means that our brains can conserve energy. This leads us neatly on to the next key point about our brains.

## 4 Brains want to conserve energy

Our brains are just two per cent of our body weight but use a huge 20 per cent of our energy (Raichle and Gusnard, 2002). The brain takes its energy from the food that we eat and the oxygen that we breathe and it burns this oxygen and glucose at about 10 times the rate of our muscles per unit mass. In fact, the brain uses up so much energy that it dies if denied oxygen for just a few minutes.

What are the implications of this?

Our brains want to conserve energy and they therefore tend to be lazy. Shane Frederick (2005) of Yale University has devised the Cognitive Reflection Test: three questions that test brains' tendency to use this lazy thinking. Take a look at one of the questions: a bat and ball cost £1.10, the bat costs £1 more than the ball. How much is the ball? Our brains want to say 10p. Is that the figure that first came into your mind? But that can't be right because then the bat would only cost 90p more than the ball. This 'instant' thinking is called System 1, sometimes also known as reflexive thinking. It's instinctive thinking that does not put too many demands on our brains. Frederick reports that more than 50 per cent of Harvard, MIT and Princeton students gave the intuitive but incorrect answer. The answer is 5p, by the way, but your System 2 or the 'reflective' system has probably worked this out by now.

## 5 Brains take 'short cuts'

This desire to conserve energy means that our brains often take the path of least resistance: they like short cuts where they don't have to work so hard. This can be useful but also problematic to us. Professor Robert Cialdini's work on influence (2001) has identified areas where we are vulnerable to influence because our brain is susceptible to taking the less demanding route. We will take a look at these short cuts in more detail in Chapter 7. These short cuts can be beneficial and a hindrance for us – beneficial in that they save effort and time and remove the need for us to think through every decision we are confronted with, but a hindrance in that we make decisions unthinkingly and sometimes these will not be the best decisions for us.

## 6 Work has changed; our brains have not

Just think about work over the last 5, 10, 20, 50, 100 years. What was it like when you started work? It has changed hugely and it will continue to do so.

Work has changed but our brains have not, and this is the challenge we have in organizations. It's an obvious fact but our brains were not designed for 21st century corporate life: they still think they are out on the savannah. Organizations need to think about how work and the working environment are structured so that they are better suited to the needs of our brains.

One part of our brain has developed compared with our ancestors and that is the part of the brain called the prefrontal cortex (PFC) – it is the part of the brain just behind your forehead. During the evolution of mammals, the PFC has evolved 3 per cent in cats, 17 per cent in chimps, but 29 per cent in humans. The PFC is very important in 21st century work: it's where we do our considered decision-making, conceptual thinking, planning and analysis. For many organizations, thinking power, creativity and collaboration are essential for success and these all depend on the PFC.

## 7 Neuroplasticity

Neuroplasticity – *neuro* from neuron and *plasticity* from plastic, meaning 'changeable, malleable, modifiable' – is one of the most useful and exciting findings to come out of neuroscience. Whereas we used to believe that once we reached 25 years of age, the brain had peaked and it was in decline all the way to old age, neuroscience has shown that the brain has the ability to continue to learn and to restructure, well into our later years. You can teach an old dog new tricks, after all. Most neuroplasticity does happen in childhood and some brain areas are better at changing and learning earlier in life – acquiring a second language, for example, and areas such as vision and hearing will tend to deteriorate for most people with age. But there is plasticity in the adult brain if we choose to learn: good news for all of us who are over the age of 25. The even better news is that plasticity in the older brain is in areas that are very important in organizations – social skills and emotional intelligence, for example – the areas that matter in leading and managing people.

Neuroscience has shown that the brain can change in response to experience, thought and mental activity. So, mental activity is not just a product of the brain but shapes it. It is the connectivity that changes in neuroplasticity: new synaptic connections form and existing synapses strengthen. The brain changes its structure with each different activity it performs, improving its circuits so that the brain becomes better suited to the task at hand and making it easier for us to perform that task. In a documentary made for the BBC, *The Human Body* (1998), Professor Robert Winston compares this learning to crossing a ravine by building a rope bridge. The first time crossing the bridge on just a rope is difficult and takes a lot of care and effort. But as the team does it again and again, building a robust rope bridge as they do so, the crossing gets easier and easier. This is a metaphor for neuroplasticity and Hebb's Law (see below): the more we do something, the easier it gets. As we repeat the task, the connection between the synapses strengthens. Think about how long it took to learn to tie shoelaces – now you would do it without thinking; or think about the first time you went to your current place of work – now you go there without having to think about the route. It is a useful metaphor to remember when we are learning a new skill and *The Human Body* is a useful film to show children when they are struggling to learn something. Learning can be hard but it gets easier with practice as the brain forges new and stronger connections between neurons.

## London cab drivers

One of the most well-known examples of neuroplasticity is evident in London cab drivers. Cab drivers have to learn the 'knowledge': thousands of roads and routes and tourist attractions across London, and it typically takes several years of intense studying to do this. Scientists at University College London (UCL) (Maguire *et al*, 2000), discovered that as the result of this learning, these cab drivers have a larger posterior hippocampus than the rest of us. The hippocampus is a part of the brain that is involved in memory and spatial awareness. Their brains restructured and grew as a result of learning so many spatial facts and mental maps of routes around London. There are many other examples too: brain scans of experienced fighter pilots show that their brains differ from those of novices – experienced pilots have a thicker cortex (outermost tissue of the brain including grey matter) and increased connectivity in the areas that manage hand and eye coordination. This increase in connectivity and thicker cortex will enable the brains of experienced pilots to have faster connections. Are you a violin player? If so, you will have denser synaptic connections to enable dexterity in your left hand. There are lots of established examples of brains that have changed as a result of learning, knowledge and practice.

## Hebb's Law

This strengthening of synaptic connections is known as Hebb's law (1949): 'cells that fire together wire together'. Neuroplasticity means that constant communication between two neurons leads them to forge a stronger connection so that they can communicate more easily together. If they can communicate more easily, they can do this faster and with less effort.

## 'Use it or lose it'

Neurons, neural systems and the brain change depending on use. Just as the synaptic connections increase and strengthen through repetition, so the connections weaken and wither if we don't use them. The brain takes a 'use it or lose it' approach: if neurons and synapses are not used, the brain prunes them away. Although people worry about losing brain cells in later life, a great deal of this pruning is done when we are toddlers and again in teenage years. The brain takes away connections that are not being used to allow space for those that are. On the whole, this is to our benefit so that the brain can allow space for those connections we regularly use.

# Some key parts of the brain – the cortical lobes

Figure 2.3 shows the cortex, which is about 2 mm thick. The cortex is divided into four lobes:

- frontal;
- parietal;
- occipital;
- temporal.

**Figure 2.3**    The cortical lobes

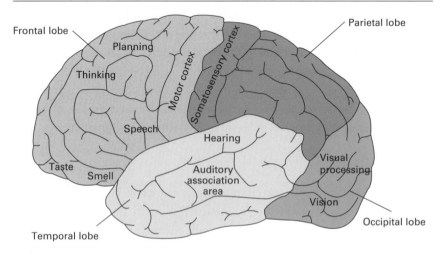

As the illustration shows, parts of the cortex have been classified, as some parts of the cortex seem to have a clear link to certain inputs and outputs from the brain. For example, the occipital lobe at the back of the head contains areas that process vision; the motor cortex has nerves that send messages down through the spinal cord to muscles. However, it is less clear cut as to what some parts of the cortex do. There are other parts of the brain that it will be useful to know about but we will cover each of these as they become relevant.

## Summary of key points from this chapter

- Our brains have grown compared to those of our ancestors 2 million years ago. One of the most compelling theories as to why they have grown is because this not only enabled us to search for food in a larger area, but also because a larger brain enables us to form stronger and more complex social relationships.
- We have about 86 billion neurons in our brains and they provide the brain with the basic units for information processing and a means of communicating between brain areas.
- But it is the number of connections, synapses, between the neurons that really make a difference to our brain's ability to work at its best.
- For our brains the goal is survival, and to achieve this they need to avoid threats and seek out rewards, but of the two responses the threat response is much stronger.
- Our brains are constantly trying to predict so that they can keep us out of harm's way.
- Our brains are 2 per cent of our body weight but use 20 per cent of our energy and are driven to try to conserve energy. They therefore use short cuts which on the whole are helpful but can hinder us.
- Good news: neuroplasticity – the ability of the brain to restructure, change and learn – can continue throughout life, if we choose to continue to learn.
- Work has changed a lot but our brains have not: our brains are not designed for 21st century corporate life and this, in part, is why we find aspects of work so difficult.

# References and further reading

Cialdini, R B (2001) *Influence: Science and practice*, Allyn & Bacon, Needham Heights MA

Dunbar, R I M and Shultz, S (2007) Evolution in the social brain, *Science*, **317** (5843), pp 1344–47

Frederick, S (2005) Cognitive reflection and decision making, *Journal of Economic Perspectives*, **19** (4), pp 25–42

Hebb, D (1949) *The Organization of Behavior*, Wiley & Sons, New York

Maguire, E A *et al* (2000) Navigation-related structural change in the hippocampi of taxi drivers, *Proceedings of the National Academy of Sciences of the United States of America*, **97** (8), pp 4398–4403

Raichle, M E and Gusnard, D A (2002) Appraising the brain's energy budget, *PNAS*, **99** (16), pp 10237–39

Silk, J B (2007) Social components of fitness in primate groups, *Science*, **317** (5843), pp 1347–51

Winston, R (1998) *The Human Body*, BBC

# Why our brains don't like organizational change    03

Think back on the world of work: what was work like when you first started out? How about for your parents – what was their experience? What did they expect from work and what did their employers expect of them? What tools and equipment did they have to do their jobs? How did people communicate with one another? Just thinking back 10 or 20 years, work looked very different and work will continue to change and the pace of that change will increase. Companies used to review their strategies only occasionally; now they have to do it every couple of years. Companies that don't change fast enough are soon out of business. Organizations constantly have to change and we expect employees to deal with it and remain resilient. Research by John Kotter (1995) and others shows that many change efforts fail and the reason for this failure is not a lack of planning but the failure of leaders to take people with them – a failure to engage them. We also know that when organizations are going through change, particularly prolonged periods of change, they are at their most vulnerable: productivity typically drops off, customer service is impaired, competitors pounce.

Helping people to accept and actively support change would be easier for leaders if they understood more about our brains and how our brains perceive change. Once they understand the brain better, they can start to work with it rather than despite it. Two key points for leaders to keep front of mind are that our brains are not designed for the 21st century workplace and they do not like organizational change. This chapter will look at why our brains don't like change and the impact of change on them. The chapters in Part Two will then focus on the practical, by exploring what we can do to help people deal with change, support it, and perform at their best.

# Why our brains don't like change

As we know from Chapter 2, our brains are prediction machines – they want to predict and make meaning. If they can predict, they are better able to keep us out of harm's way. What does significant organizational change mean? It means our brains cannot predict what is going to happen. Ambiguity is even worse because our brains really don't know what to make of it. In fact, research studies suggest that we can live more comfortably with certainty about a negative outcome than uncertainty. Huntington's Disease is traumatic: it causes physical decline and brain damage when people reach their thirties and it gets progressively worse over time. As yet, there's no cure and its progress can't be reversed or slowed down. It causes problems in movement, mood swings and personality changes. Eventually people need full-time nursing and suffer an early death. It is genetically inherited: if one parent has the disease, then children have a 50 per cent chance of having the gene. To know that you have this disease would be devastating. However, research (Wiggins *et al*, 1992) suggests that people who have the test, whether the test proves negative or positive, feel better – that is to say, lower scores for depression and higher scores for well-being – than children of sufferers who live with the uncertainty of not knowing whether they have inherited the disease. Once we have certainty, we begin to know what the story is, we can make meaning, tell ourselves a narrative about what is happening to us, we can begin to plan, based on that story. Once we have meaning, we spend less time trying to make sense of it, less time and energy ruminating. But if we don't have certainty, we don't know what the story is, and we endlessly work through the different possibilities and the various scenarios and the potential impact on us.

I saw this in one of the large UK banks after the global banking crisis of 2008: the bank pulled together those parts of the company that were going to be closed down or sold off and a significant number of employees found themselves in this division or volunteered to join it. You would imagine this would be an unhappy and anxious place to be, with employees worried about the future. However, employee engagement surveys showed the employees in this part of the bank to be more positive than those in the main bank. When I asked employees about these findings, they put them down to the fact that although they knew they would have to leave the bank at some point in the near future, at least they knew this for a fact and could start to plan around it. They had a narrative they could work with. Those who had volunteered to be in this part of the bank were noticeably more upbeat, because they had chosen this path. Although employees in the main bank seemed to have more certainty about their future, because of the precariousness of any job in the banking sector at that time, they actually felt more anxious because they were less sure about their future than those in the division being closed down.

The ability to predict is not the only reason why our brains like certainty. Being certain removes psychological discomfort. When we are in a familiar and comfortable situation at work, our brains can work on autopilot, so to speak, using neural pathways that have been used many times before, with strong synaptic connections. However, when we feel uncertain, several areas in the brain become active, in particular those parts of the brain that are part of our fear network. Uncertainty registers as an error or a gap – something that must be dealt with before you can feel comfortable again. The orbitofrontal cortex (OFC), part of the prefrontal cortex just behind our foreheads, appears to have a role in reporting whether our current situation is dangerous or worrying in some way and in guiding our decisions and behaviour appropriately (people with damaged OFCs cannot properly weigh up the chances of failure). The dorsal anterior cingulate cortex (dACC) acts as an error detector and is part of the brain's pain network. When we detect a conflict, activity in the dACC increases (Botvinick *et al*, 2004) and drives us to adjust what we are doing to resolve the conflict. When we are feeling uncertain, there is more connectivity between the dACC, the amygdala – which deals with emotions and is particularly associated with fear – and the insula, which generates social emotions such as pride and guilt. The insula also helps us to prepare ourselves, eg helping us to imagine what a cold swimming pool will feel like before we jump in. Research using fMRI (Hsu *et al*, 2005) shows that the more ambiguity we are faced with when making a decision, the more activity there is in the amygdala and orbitofrontal cortex and the less there is in the striatal system – part of the reward network. It follows that having certainty is rewarding to our brains: it is linked to the activation of the ventral striatum and release of the neurotransmitter dopamine. This in turn feels good and has a positive impact on our ability to focus and perform. Think about a time you were confused about something, and then someone clarified what they meant or what was going on: you then feel much better.

There is an important message in all of this for both leaders and employees: if people are struggling with change, it is not a sign that they are weak or failing in some way. Our brains don't like it. They are prediction machines that are not designed to deal with the ever-increasing speed and quantity of organizational change. Our brains don't feel comfortable with it. We need to be empathetic towards ourselves and towards others. That is not to say be complacent – the rest of this book provides thoughts and strategies as to how to help people adapt to, and support, change.

**Figure 3.1**   Critical systems in dealing with uncertainty

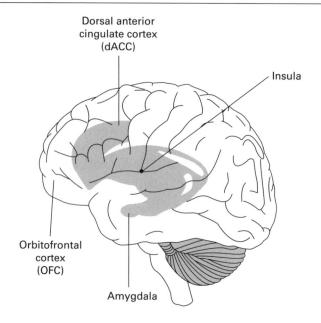

# The impact of change on our brains

Mild uncertainty, something slightly out of the usual, can be quite appealing to our brains: they like a little bit of novelty, and in fact, it is beneficial to us (we will come back to this in Chapter 4). But significant organizational change or layers and layers of small changes are seen by our brains as a threat. The threat response is well known in the form of the 'fight or flight' response (and there are two more related responses: flock and freeze where animals huddle together for safety or become motionless with the hope of not being noticed). Fight or flight means that the brain and body get ready to deal with the threat and send energy to those parts of our anatomy that are crucial in preserving us. Triggered by the amygdala, our hearts start to pound, digestion slows, the immune system is suppressed, cortisol (the stress hormone) increases, and the blood vessels to muscles dilate.

As this happens, blood flows away from the PFC (Bishop *et al*, 2004), where we do our considered thinking and planning and manage our emotions, and goes to those parts of the brain that get us ready to fight or run – see Figure 3.2. As blood flows away from our PFC we cannot think as well or as clearly. The aroused amygdala means that we become anxious and start to see threats in the workplace as greater than they really are and we start to see threats where they don't even exist. Uncertainty distorts our view of threats and can make them seem even worse. When we are in the midst of uncertainty, we are more likely to expect the worst and this makes uncertainty even more stressful. Uncertainty about the nature of a negative event is far more stressful than certainty. For a child, for example, of the two punishments 'Go to your room and don't come down again this evening' and 'Go to your room and just wait until your father gets home...', the latter is much more frightening because of the unknown nature of the punishment.

**Figure 3.2**   The impact of change on our brains

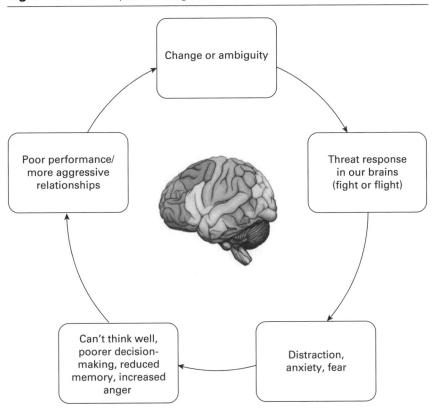

Uncertainty is stressful in the workplace too. Think of a time when you were facing a great deal of uncertainty and change at work. Imagine you walk into the office one morning and you can see that all your colleagues have been invited to a meeting but you have not: how does that make you feel? Or your boss arrives and normally she is friendly and says 'hello' but today she looks distracted and does not want to make eye contact with you. How do you feel about that? The threat response means that we begin to worry and we begin to think about the adverse possibilities. We speculate and we speculate negatively.

Decision-making becomes more difficult because we can't think clearly; sometimes even the simplest of decisions become difficult: what do I want for lunch? What kind of coffee would I like? Our brains feel overwhelmed and fuzzy. As the threat response continues, we begin to see colleagues as a threat (this takes us back to our ancestors – if resources were under threat then we might begin to see other tribe members as competition for those resources) and we are less able to collaborate. Our memory and ability to process new ideas and information is reduced, and we are less able to focus on the present. It impairs our thinking, creativity and problem-solving. It becomes difficult to focus. Our field of vision literally narrows – this too goes back to the brain's instinct for survival. If the sabre-toothed tiger suddenly appeared, your heart would pound, your muscles would tense and your vision would narrow to focus on the tiger: this improved your chances of survival.

We are less resistant to stress and to emotions. In fact, the adult brain in a threat response is much like that of a teenager – quick to get angry and emotional, hard to reason with. So an organization going through change is like an organization being run by a group of teenagers. Then it gets worse: as we realize that we are distracted and can't think straight, we become aware that we can't work at our best, the threat response increases, and down into a negative spiral we go.

Just when organizations need people to be thinking at their best – to implement change, manage integration, learn new skills, use that new IT system, set up new teams, continue to focus on the customer – their minds are distracted and their thinking is impaired.

There are longer-term implications too. For our ancestors in the wild, sudden bursts of cortisol were beneficial because the hormone helped them to survive. Once the predator was no longer a threat, cortisol levels would drop. Robert Sapolsky (1998) of Stanford University contrasts the benefits of cortisol on animals in the wild and the impact of cortisol on people in the modern world – the impact of man-made stress. We have created work environments where people are frequently under stress and cortisol is constantly in the system. As Sapolsky puts it, no zebra running away from a lion would understand why human beings secrete the same stress hormones out of fear of, for example, public speaking. The response to the stress becomes more damaging to humans than the original stressor itself. In the long term, cortisol is damaging physically and mentally. It is particularly detrimental to the hippocampus which plays an important part in memory (Lupien *et al*, 1998).

Of course, the impact of change and people's levels of resilience vary. Some are much more able to cope with uncertainty. In particular, people who have been through change in the past and who have come through it successfully can (consciously or subconsciously) apply that previous experience to the current situation. Past experience has a major impact on how we perceive the present: our brains are constantly subconsciously comparing the present with our past experience. If the past experience was satisfactory or pleasant, then we are more likely to feel OK about the current situation, but if the past experience was painful, these emotions come flooding back and colour how we experience the current situation. In recent conversations with employees in a government department, people talked about their experience of a restructure that had happened five years ago. They are now faced with a new restructure. Some of them talked about the fact that even though they had got the role they wanted from the last restructure, they felt anxious about the imminent changes. Although the eventual outcome had been positive for them, the process of having to reapply for their jobs had been very uncomfortable. The new restructure reawakens those painful emotions from five years ago. This is one of the reasons why people's reaction to change can vary so much and be unpredictable: past experiences influence how they feel. Past experience and expectations shape how we experience the present.

# Summary of key points from this chapter

- As part of the drive for survival, our brains want to be able to predict and make sense of the world. Organizational change takes away this ability to predict and sends our brains into a threat response.

- Research studies show that we are more comfortable with certainty about a negative outcome than we are with uncertainty.

- The threat response means that blood flows away from our prefrontal cortex, which is important for thinking, decision-making and emotional control, and goes to those parts of the brain that get us ready for 'fight or flight'.

- The threat response means that our thinking is impaired, we have less control over our emotions, we see the workplace as more hostile than it really is and our performance drops off.

- Past experience of change has a major influence on how we perceive current changes. Expectations colour how we see the world.

# References and further reading

Bishop, S *et al* (2004) Prefrontal cortical function and anxiety: controlling attention to threat-related stimuli, *Nature Neuroscience*, 7 (2), pp 184–88

Botvinick, M M, Cohen, J D and Carter, C S (2004) Conflict monitoring and anterior cingulate cortex: an update, *Trends in Cognitive Sciences*, 8 (12), pp 539–46

Hsu, M *et al* (2005) Neural systems responding to degrees of uncertainty in human decision-making, *Science*, 310 (5754), pp 1680–83

Kotter, J (1995) *Leading Change: Why Transformation Efforts Fail*, Harvard Business Review

Lupien, S J *et al* (1998) Cortisol levels during human aging predict hippocampal atrophy and memory deficits, *Nature Neuroscience*, 1 (69–73)

Sapolsky, R M (1998) *Why Zebras Don't Get Ulcers*, WH Freeman & Co, New York

Wiggins, S *et al* (1992) The psychological consequence of predictive testing for Huntington's disease, *New England Journal of Medicine*, 327 (20), pp 1401–05

# PART TWO
# What can we do?

**P**art One of this book has set the context – why organizations need to pay attention to neuroscience and how every leader and every employee will benefit from understanding the basics about how their brains work. Chapter 2 described some of the understanding we need to have about our brains and some of the principles that guide them, and Chapter 3 explained why our brains don't like significant change and the impact change has on our ability to think.

Part Two of this book moves on to the practical – with an understanding of the brain and drawing on research studies – what can we do to enable people to perform at their best during organizational change? Each chapter takes a look at the relevant science and research, provides examples of practical things organizations and leaders can do, and ends with a summary of key points and a series of questions to help plan.

# Performing at our best during change

04

For many organizations, employees' brainpower is all they have. It's the difference between success and failure, innovation and stagnation. We all have moments where we are focused and know our brains are working at their best; equally we all have bad days where we just can't quite get on with our work, either because we feel lacking in motivation or because we are feeling anxious or stressed. It gets to the end of the working day and we wonder where the day has gone and what we achieved. What makes the difference? The challenge for all of us, but particularly for leaders, is how to counteract this unproductive, distracted state. If change is inevitable, what can we do to keep ourselves and other people performing at or near their best? This chapter shares the insights from neuroscience together with practical interventions that leaders have successfully used.

## The science

### *The impact of threat and reward states*

Chapters 2 and Chapter 3 described the threat and reward states in our brains and explored the negative impact of the threat state on our ability to think and collaborate at work. Figure 4.1 is a summary of the impact of the two states on our brains and on our ability to think and perform. On the left-hand side, the threat state sends our brains into a distracted and anxious state that is also physically damaging in the longer term. But take a look at the right-hand side. There are many different definitions of employee engagement but, to me, the description of the brain in a 'toward' state is the description of an employee who is engaged in their work – positive, open to learning and willing to pitch in with others. Leaders need to keep these two states front of mind and be aware of the consequences of what they say and do. Our brains, as set out in Chapter 2 and Chapter 3, are much more sensitive to the threat state and so are easily pushed to the left. Organizational change and uncertainty will compound this problem.

**Figure 4.1**    The impact of threat and reward on our brains and on our ability to think and perform

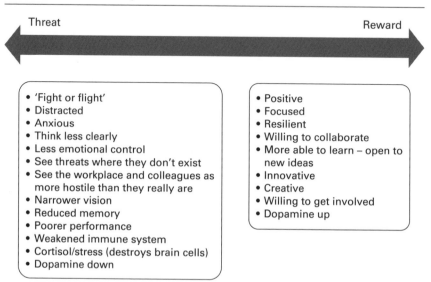

The inverted U of performance is a 'law' that dates back to 1908, developed

## The inverted U of performance

The inverted U of performance is a 'law' that dates back to 1908, developed by two psychologists – Robert Yerkes and John Dodson – but that has stood the test of time and has been investigated further in more recent years. Figure 4.2 shows a simple graph of the inverted U. On the y-axis is the brain's ability to stay focused: at the bottom of this axis, the brain is disorganized, forgetful and distracted; at the top of the axis the brain is organized and focused. Running along the x-axis is the level of stress the brain is under. At the top of the inverted U, the brain is sharp and performing well. On the left-hand side of the inverted U the brain is unfocused, and over on the right-hand side it is not performing well either: it is distracted. The Yerkes–Dodson law states that there is a relationship between the level of arousal and our ability to perform a task, ie there is an optimal level of arousal: too much or too little arousal reduces our ability to perform the task well.

**Figure 4.2** Inverted U of performance: Yerkes–Dodson

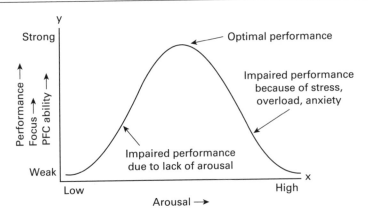

We have all had those days when we are at the top of the inverted U: we are writing a document and the words just flow; we are making a presentation and we are articulate and clear and we can see we are holding people's attention; we are writing a proposal and the arguments for what we are proposing come to us easily and we know they are exciting and persuasive. Performing these tasks feels almost effortless: our brains feel vibrant and focused. But equally we have all had those days when we can't get motivated to do what we need to do. Perhaps we have a deadline for some work but that's not until Friday and today it's only Monday: we tell ourselves that we will start to write the document at 11am but we drift off to make a cup of coffee, we start to read old e-mails, we check the Twitterfeed; if we are working from home, we start to tidy the house. We can't quite bring ourselves to knuckle down and get on with the work. We get to the end of the day, and we really haven't achieved as much as we could have done. This is the consequence of too little arousal, too little stress.

On the right of the x-axis are the days when we can't think straight because we feel overwhelmed with information, tasks and deadlines. We have a report to get in by lunchtime, but there's another internal client who says they urgently need some information from us before then, the e-mails keep popping up, the phone won't stop, we also have another pressing deadline for tomorrow. We know the next three months look pretty similar so there's no let-up. Perhaps we have a frail parent whom we know is going to need more support from us and we have a colleague who is being decidedly difficult to work with right now and another one who has gone on sick leave just when we really needed her to be around. Our brains feel frazzled. We don't know where to begin with all these tasks. Things feel out of control and overwhelming.

The inverted U sets out what we have all experienced: under-stimulation and too little pressure lead to us being easily distracted; too much pressure and our brains find it hard to think. Amy Arnsten of Yale has investigated the inverted U further by looking at the levels of catecholamines (chemicals such as dopamine and norepinephrine) in our brains in these different states on the inverted U (Arnsten, 2009). Norepinephrine strengthens the relevant prefrontal cortex (PFC) network connectivity, whereas dopamine helps to keep us focused and to block distractions from our goals. Both need to be in just the right balance to keep our PFC functioning well: the right level means that we can focus and move towards our goals; too little and we are apathetic and distracted; too much and we become stressed and the PFC shuts down and this impairs our decision-making ability. In fact, the PFC is sometimes referred to as the 'Goldilocks' of the brain as the chemical balance has to be just right for it to perform optimally. This matters to each of us as individuals and to organizations as a whole, because the PFC is a particularly important part of the brain for succeeding at work, and in life generally. It is the network of neurons that guides our attention, thinking, planning, actions and emotions. By combining external stimuli and ideas stored in our memory, it helps us to make meaning and form perceptions. Twenty-first century organizations need this part of their employees' brains to be working at its best as often as possible.

Many organizations could learn a lot by taking a look at the inverted U. In recent years, in both the public and private sector, there has been a belief that if we just push people a bit harder, we will get a bit more out of them. Some leaders believe that this is a 'strong' way to lead and that it shows a commercial toughness that will push performance on. The inverted U acts as a warning to such thinking. People do need some pressure, but finding the optimum amount of pressure is hard to do, and if we push people too hard their brains start to shut down. Too many leaders and organizations think they should push people harder to get more out of them, but this is actually counterproductive.

## Flow

The concept of flow, as Csikszentmihalyi (1990), the Hungarian-born positive psychologist puts it, is a similar way of describing being at the top of the inverted U. Being 'in the zone' is another description. Csikszentmihalyi called it flow because the feeling is that whatever you are focused on comes easily. You feel fully immersed in what you are doing: you are focused and feel motivated to achieve the task in hand. It can come during sports activities – professional tennis players talk about seeing the ball as if it is much larger than it really is, because hitting it becomes so much easier when in flow – and it can come when one is occupied with a solitary activity, playing music or writing. The key criteria to flow are (Nakamura and Csikszentmihalyi, 2001):

- A clear challenge that captures your attention.
- You have the skills to meet the challenge.
- You get immediate feedback on how you are doing.
- You get a positive feeling from what you are doing.

So, as in the inverted U, it's the state where your brain reaches just the right balance in terms of challenge and ability to achieve the task.

## *Dopamine*

Dopamine is a neurotransmitter – a chemical that is responsible for transmitting signals between neurons in the brain. Few neurons create dopamine and those that do are largely in the substantia nigra and the ventral tegmental area (VTA) in the midbrain. It is released to various parts of the brain including the prefrontal cortex and impacts brain function in many different ways, including cognition, voluntary movement, motivation, mood, working memory and learning. People who suffer from Parkinson's Disease, for example, lack dopamine and this leads to their difficulty in movement. The right levels of dopamine are critical for focusing and attention, and dopamine is also thought to play a part in Attention Deficit Hyperactivity Disorder where children, amongst other problems, tend to have short attention spans, find it hard to concentrate and struggle to carry out instructions. Drugs that can increase dopamine in the brain such as Ritalin, Modafinil and Concerta are prescribed to help lessen these symptoms. There are lively debates in academic circles and in the media about the rights and wrongs of people with healthy brains taking such drugs to enhance their brain's performance and concentration. Professor Barbara Sahakian of Cambridge University is known for her work on early detection of neuropsychiatric disorders and also for her work on cognitive enhancement: she writes and speaks frequently on these subjects and on the ethics of cognitive-enhancing drugs (Sahakian, 2014).

Amongst its other roles, dopamine is also believed to provide a teaching signal to the parts of the brain that enable us to acquire new behaviours. Dopamine increases the number of new connections made in the brain and has been shown to reinforce the link between a stimulus and reward (Wise, 2004) that is important in goal-directed behaviour (see more on this in Chapter 10). Dopamine is associated with reward and in particular with expectation of reward. Schultz *et al*'s work on 'prediction error' (1997) shows that the firing of dopamine changes when animals receive unexpected rewards, and when they expect them but are disappointed. When an animal receives an unexpected reward, dopamine firing increases. When an animal experiences a stimulus that suggests a reward is on its way, dopamine is fired in response to this stimulus, but the animal has no increase in dopamine firing when the expected reward actually occurs. If the animal receives the stimulus that predicts a reward, dopamine increases in response to the stimulus; but if the animal is disappointed and no reward appears, dopamine firing is suppressed. There are obvious negative implications here for organizations when employees' expectations are not met, and positive implications when they receive unexpected recognition or rewards (and these do not need to be monetary, as we will see in Chapter 5 on our social brains).

Both human and animal brains like dopamine and so we tend to seek out ways in which we can increase our dopamine levels. In the 1950s James Olds and Peter Milner discovered by accident (they had intended to research sleep), that if a rat had certain parts of the brain stimulated, it would repeat the required behaviour to get further stimulation as it was so pleasurable to it. Dopamine was the chemical being generated. In later experiments, rats were found to press a lever over and over again to get this stimulation and would even go without food or rest. Dopamine levels rise when we experience something new or good or when we want to do something. Our brains like dopamine and will seek it out.

Lack of dopamine causes a range of problems. Dopamine levels can suddenly drop if we are disappointed, as mentioned in Schultz *et al*'s research. A reduced dopamine level reduces our desire to do anything significant and has a negative impact on our ability to think creatively. Low dopamine levels lower mental performance.

## Mindset – the power of believing you can improve

Take a look at the following four statements – what's your view?

1 Your intelligence is something very basic about you that you can't change very much.

2 You can learn new things, but you can't really change how intelligent you are.

**3** No matter how much intelligence you have, you can always change it quite a bit.

**4** You can always substantially change how intelligent you are.

These four statements come from *Mindset: The New Psychology of Success* (2007) by Carol Dweck, Professor of Psychology at Stanford University. Dweck has spent decades researching achievement and success. Her research has revealed a simple but very powerful idea: the difference made by believing that you can improve. The four statements are a simple way of testing our own beliefs. If you agreed with the first two statements, that suggests a 'fixed' mindset where you believe that intelligence is something you are either born with or not and there's not much you can do about it. If you agreed with the second two statements, that suggests a 'growth' mindset: a belief that if we try, we can learn and develop.

When we have a fixed mindset about something, we believe that basic abilities are fixed traits. We have a certain amount of ability and there's nothing we can do about it. That's that. The best tactic is therefore to try to look as smart as you can whenever you can. Don't put yourself in a position that might make you look stupid. By contrast, when we have a growth mindset about a skill, we believe that we can get better at it through practice, persistence and effort. It's not that everyone has the potential to be brilliant at everything, but we can get better if we try.

We all have fixed mindsets about something. In her book, Dweck acknowledges that she had a fixed mindset in her youth. The educational system often reinforces this view and I think many of us emerge from school and university thinking this way. There's a tendency more generally to look down on people whom we believe have to 'put in the hours', tending to dismiss them as being people who have to work hard to get anywhere. The underlying assumption is that if you are clever you shouldn't have to work that hard. We tend to admire what we see as natural, effortless achievement.

Mindset does not just relate to intelligence either. It's relevant in many areas of life. We tell ourselves, I'm no good at art, I can't ice-skate, I'm a lousy cook. We all have an area where we have a fixed mindset. On reading Dweck's book, I realized that, as an eight-year-old, I gave up the piano because I told myself that my sisters were the musical ones, not me.

But what about the growth mindset? Where are the good examples? The world of sport provides many: Michael Jordan in a Nike ad talks about the numbers of failures he has had – with the suggestion that these spurred him on to try harder (see p 56 for online reference). Jonny Wilkinson talked about being obsessive about rugby practice and believed it gave him the extra 1 per cent. David Beckham says his secret to footballing success was practice. The top tennis players talk about the dedication their sport has required from them from an early age and, when they lose a match, talk about what they have learned, what they will work on, the extra effort they will put in on the court or in the gym. This is the growth mindset. It's not that everyone is going to be a great athlete – some are more gifted – but we can improve, and that's the key point to remember.

Why should organizations care? During organizational change, people will inevitably face difficulties and setbacks. How people deal with these problems, and whether they are able to deal with the causes of the problems, will be a key determinant of whether the change programme will be successful. Dweck and others have carried out many studies over the years that provide overwhelming evidence that people who have a growth mindset will try harder and therefore, in both the short term and the longer term, will do better (Blackwell *et al*, 2007). Much of Dweck's work has been done in schools but it applies equally in organizations. If fixed mindset people do badly at something, they believe that their intellect has been judged and they have failed (Nussbaum and Dweck, 2008). Because they believe that failure shows that they are not that bright after all and there is nothing they can do about this, they tend to run from difficulty, challenge themselves less, take on easier tasks that will make them feel OK about themselves again. There's also a risk that people with a fixed mindset prefer to surround themselves with people who are less clever than they are, as they will make them look and feel good.

In a research study conducted in Hong Kong (Hong *et al*, 1999), students who had performed poorly on their English proficiency exams (a subject that is essential because all classes are taught in English) were less likely to participate in additional English lessons if they had a fixed mindset than those who had a growth mindset.

In organizations the implications of this are obvious: in organizations where employees have a fixed mindset they will seek situations that validate them and make them look good. Having made a mistake, rather than being persistent and trying to learn from the error, they will withdraw to their comfort zone. Leaders and managers with a fixed mindset will believe that future great managers are 'born' and will put less effort into developing people: you have either got talent or you have not. A college friend of mine joined the graduate programme of one of the large oil companies: at the tender age of 23 he was told how high, or not, he would rise in the organization by the end of his anticipated 40-year career there. Not surprisingly, he left this fixed-mindset company.

The additional danger is that fixed mindset leaders tend to surround themselves with people who make them feel clever and who do not challenge them. In his book *Good to Great* (2001), Jim Collins identifies one of the key criteria that distinguishes long-term successful companies from their less successful competitors as their willingness to 'confront the brutal facts', ie successful companies are willing to look at mistakes and learn from them, while remaining confident about their ability to succeed. This reflects a growth mindset: recognizing that making a mistake can be a useful thing to do if you learn from it. Fixed mindset people have a need for constant validation and will not admit to, or dwell on, their errors.

Neuroscientists have studied how the brains of fixed and growth mindset students differ when they have made a mistake (Moser *et al*, 2011). The researchers measured event-related potentials (ERP) – electrical brain signals caused by external or internal events. People with a growth mindset had more activity in their brains in response to making an error and this enhanced their accuracy in trials after making mistakes compared with people who had a fixed mindset. This reflects the difference that can be seen in their behaviour. Fixed and growth mindset people have different reactions to failure: fixed people don't want to deal with it; growth mindset people see it as a challenge and want to learn from it.

What can we do? One of the great things about mindset is how easily it can be changed (in fact in some of the experiments referred to above, participants were deliberately put into either a fixed or growth mindset). Just teaching people about the concept and the difference the two mindsets make, creates a difference to performance (Aronson, Fried and Good, 2002). Teaching people about neuroplasticity – that we can learn and improve, that we can develop and strengthen the connections between neurons – makes a significant difference. You *can* teach an old dog new tricks.

## When do you have those 'aha!' moments?

Think back over the last few weeks: when have you had one of those eureka moments, when you have been struggling with an issue and suddenly the solution comes to you? Going for a walk, driving the car on a familiar route, in the shower, daydreaming, lying in bed? It might well have been in one of these settings, but I have not yet heard anyone say that it was sitting at their desk, staring at their laptop. There's a biological reason for this: when you are in a meeting, working on a document, your brain is focused on doing those tasks and it has a limited capacity. But when your brain is relaxed it can run free, so to speak, making connections across the brain and coming up with a new idea. The brain reaches the alpha range of waves, 8–13 Hz (these can be observed through EEG caps): the brain is 'quiet' and can roam around, searching for connections and ideas. This is what happens just before the moment of insight (Fink and Benedek, 2014).

One way for neuroscientists to explore moments of insight in the lab is to give people word puzzles to solve. The word puzzles are the kind where there are three words and the participant has to think of a word that works with them all to make another word, eg

Pine –

Crab –

– sauce

(the answer is at the end of the chapter, in case you want to have a go at solving the puzzle).

Participants are then asked to tell the researchers how they solved the problem – did the idea just come to them as in an 'aha' moment of insight where the answer seems to come from nowhere, or did they methodically work their way through possible solutions (or did the answer not come to them at all)? The scientists can see from the pattern of brainwaves, just *before* the participant comes up with the answer, which process their brain has used. For a moment of insight to happen, the brain needs to quieten down, and just before the idea comes, alpha waves appear. Certain areas of the brain, such as memory, are activated. These same areas are activated during sleep, hence the fact that good ideas often come to us in our sleep, or perhaps more accurately during what is known as hypnagogia – that slightly disoriented state between waking and sleeping – which is also associated with lower frequency alpha and theta oscillations (Gruzelier, 2009).

## What can we do? Solutions and examples of what other leaders have done

Typically, when an organization goes through significant change or there are layers on layers of change, performance drops. Occasionally, people's drive can increase. I have seen groups of employees where, in a fighting spirit of defiance, their productivity has increased, as if to show that, despite redundancy being on the cards, they will show the world just how good they are at their jobs. But for most organizations most of the time, the challenge is how to keep people focused and able to deliver what's expected of them.

Based on what we understand about the brain and motivation, what are some of the things that organizations can do to keep people focused and performing at their best during change? At one level neuroscience is complex, but at another level applied neuroscience can be very practical and straightforward. This section looks at the useful things leaders and managers can do to keep their people performing well. Some of them will be familiar: one of the appealing aspects of neuroscience is that it reinforces people's gut instincts as to how to get the best out of people. It gives confidence to leaders when they are trying to decide whether a certain course of action is a wise one. In workshops, very experienced leaders have realized that some of the activities described below were not frivolous or indulgent, but actually helped to put people into a more constructive and fruitful frame of mind. Neuroscience shows that they are not just nice to do but will enhance performance. The other great benefit of the suggestions below is that every leader in the organization can just get on and do them straight away. They don't have to wait for colleagues to change or for the culture to transform – they can take the initiative. Another advantage is that they do not have to cost money – these are small things that, together, can make a big difference.

## Set short-term goals that people can achieve

If people are struggling and have long-term objectives that stretch out over months or even years, break those objectives down into shorter-term goals. Achieving a goal generates dopamine in our brains and makes us feel good. Dopamine also helps us to be focused and to have a positive state of mind. So, providing employees with goals that they can achieve today or this week will make them feel good and put them into a positive frame of mind where they can go on to tackle the next task. The good feeling created by achieving short-term goals is probably why many people like to have checklists where they can cross things off when they are done. As some people confess, they even add a task to the list after it has been done, because they can then cross it off and this makes them feel better! Note: There are times when focusing on longer-term goals is beneficial. For more on this, see Chapter 10.

## Remind people of past achievements

Reminding people of past achievements can generate reward-related dopamine and activates the same parts of the brain as if the person were experiencing that achievement right now. So, if someone is struggling with change or feeling unsure of their ability to learn new skills or deal with an unfamiliar situation, a simple and easy thing to do is to ask them to talk about the last time they did something really well, or something they are proud of having achieved. It can work on a one-to-one basis or in a team meeting. I often do this at the beginning of workshops with teams, as I know it will get them in a more positive state of mind, generate dopamine and in turn that will mean that they are more open to learning. It is a technique that works well with a team that is being disbanded and about to go into new roles. With a senior management team in part of a rail company, I asked them to list all the things they felt they had achieved as a team and as an organization. The exercise helped them to recognize just how much they had achieved. As Bridges acknowledges in his book *Managing Transitions* (1991), such exercises help people to let go of the past and help them adapt more quickly to the next role. It treats what people have achieved in the past with respect but also reminds them how capable they are.

## Winning breeds winning – it's a virtuous circle

On a different but related note to remembering past achievements, another way to boost performance is to enable people to do something well: winning really is a habit. It changes our hormones and our mindset. Once people have done one task well, their confidence will be boosted and they will be better placed to take on the next challenge. So, how can you enable someone to win or succeed at something?

## Give praise and recognition

We all know that we should give people praise and recognition where it is due, but neuroscience teaches us that it's not only nice to do but it will also help to keep people performing. That said, praise needs to be genuine and specific: in some organizations there is a culture of too much praise – people expecting to be praised and almost in constant need of validation (perhaps a reflection of a 'fixed mindset' mentality). Praise is more useful to people when they know why what they did was good so that they know what behaviours to repeat: we need to learn to praise wisely. Praise and recognition make us feel good because they generate dopamine and help to put us into a positive 'toward' state which will enable us to perform that bit better on the next task. Unexpected rewards generate even more dopamine and so are even better: the bunch of flowers you receive and that you were expecting is lovely, but the bunch of flowers that arrives out of the blue is even better.

There are many ways of recognizing people at work, and social media in the workplace has made it much easier for employees to thank each other. One manager said he sent out a weekly e-mail summarizing what had been done well this week – it was good for him to pause and recognize what his team had achieved, nice for them to receive the e-mail and see that their good work had not gone unnoticed, and the positive e-mail set up the team well for the following week.

## Perform an act of kindness

Research by Chancellor *et al* (2018) explored the impact of asking a group of people in a Spanish company to perform five kind acts over the course of four weeks. Both the receivers and the givers of the kind acts reported improvements in their well-being in the short term and in the long term, eg two months later receivers were happier and givers were less depressed and more satisfied with their work and lives, compared with a control group. In addition, receivers were more likely to perform small acts of kindness to others. Small acts of kindness can be contagious in the workplace.

## Novelty

Our brains might not like lots and lots of change but they do like a bit of novelty. To the brain, too much uncertainty is dangerous, but the excitement the brain gets from novelty keeps us motivated to find new and better environments and resources. It's easy to see why this might be important in terms of evolution. As we have all experienced, one of the challenges for the brain is boredom: our brains are constantly having to fight off distractions to help us stay focused on the task in hand. Working on the same machine all day or doing the same repetitive task at the desk means that our attention will begin to wander and we will become distracted and make mistakes. Sitting in the open-plan office having to write a rather dull document, we will be tempted to tune in to conversations around us, make cups of coffee; sitting in the same team meeting week in week out with the same structure and where we can almost predict who will say what, our brains begin to tune out. We're bored. Our brains are on the left-hand side of that inverted U. So, our brains like a bit of novelty, something to pique their interest.

The novelty you choose to introduce might be large or small: leaders and managers have reported having success from being a bit more creative with the titles of the e-mails (when people are bombarded with so many e-mails each day, how can you get recipients to want to read yours?), changing the format and location of the team meeting (just a different location can help people to look at issues slightly differently). One bank has changed the nature of its team meetings. Leaders and managers have learned to go into meetings with no agenda but just to listen to employees. This was daunting at first for some managers, but the bank is now seeing very useful results from it – employees can talk about the real issues on their mind, managers learn to reconnect with their teams and, from the feedback, leaders understand what is really getting in the way of people being able to do their jobs.

But don't overdo it in terms of novelty: if every day something is new and different, then our brains begin to struggle – too much novelty and they can't predict and go into threat mode.

## *Laughter*

Sometimes there can be a 'gallows humour' when going through major change; sometimes there is just an instinct that we need to lighten the atmosphere. It is striking how many teams, no matter what they are faced with, still find time to have a bit of fun together: a light-hearted quiz on a Friday afternoon, a team 'bake-off' where people bring in cakes they have made, a silly poem competition. Neuroscience also shows that laughter is good for our brains. In some research conducted by Warwick University (Oswald *et al*, 2009), the researchers demonstrated that making people happy (either by making them laugh by watching a comedy DVD or by giving them chocolate or fruit) improved their productivity by around 12 per cent. In this particular study, before watching the DVD, participants had to add up a series of five two-digit numbers. They were also under pressure as they were given just 10 minutes to complete the task. Across the entire sample, participants answered just under 20 additions in 10 minutes. Having watched the DVD, the happy participants improved by around two correct answers, an increase of 10–12 per cent. With another group, participants were asked to talk about an unhappy event before doing the second set of tests: people who had just recalled an unhappy memory performed worse in the second productivity task than they had in the first. Not that having a bit of fun on its own will be enough, as many longitudinal studies, neuroscience and our own personal experience demonstrate: we need to feel fundamentally engaged at work to be productive, but a bit of laughter (or chocolate) can help.

## *Provide information*

Our brains crave information, and this makes sense since, as we saw in Chapter 2, our brains' goal is survival and being able to predict puts our brains in a better place to protect us. The more information our brains have, the more accurate the prediction is likely to be. Our brains like certainty: if we don't have it we start to speculate, to fill the gap. Not having certainty uses a huge amount of our energy and is very distracting. We see it in children from an early age: which one of us has not sat in the back of the car and asked, 'Are we nearly there yet?'. Think of the difference it makes to us when we are standing on a train platform and we have the indicator board telling us that our next train is due in 10 minutes – remember how it feels to be on a platform with no information about when our train will arrive (as used to happen at my local station) – will it be two minutes or could it be 30 minutes? How long will I be here? Am I going to be on time for my meeting or will I be late? We become slightly more anxious. This might seem like a small piece of information but our brains are much more settled when we have it. The same is true in the workplace, especially when we are going through change. Information is rewarding to people's brains: if you choose to hold information back, people will be distracted and will be filling the gaps with their own gossip and rumours. What's more, employees' brains will be in a threat state (see Chapter 3) and so will speculate negatively and see the workplace as a more threatening place than it really is. So, if leaders are hesitating to communicate, they need to be aware how distracted employees' brains will be as a consequence.

Often it is difficult to know when best to communicate, especially as things can shift so quickly during change and there are not many certainties. If you communicate a lot early in the process, you might find yourselves having to explain why plans have changed. There is no perfect answer here: there are advantages and disadvantages to communicating early, but organizations need to be aware of the energy employees will expend by not knowing. Many leaders and managers also report that when there is a lack of information their teams tend to assume that they must know more than they are saying and that they are withholding information (the threat response at work again).

To create some certainty for employees during change, one thing organizations can do is at least be clear about the communication process, eg that you will commit to update all employees every Thursday morning, even if there is no news. This way, employees can be certain of the process, even if not the content or the impact of the changes in the organization.

If tasks that are being given to employees are complex, it can be useful to break them down into a subset of tasks. This also provides employees with more certainty, and has the benefit, as mentioned above, that they can achieve a sub-task more quickly and this will get them into the virtuous cycle that dopamine can create.

## Allow people to reach their own insights

While the brain craves information, it also wants to reach its own insights. The importance of allowing employees to reach their own insights is something that organizations hugely underestimate. Typically in change programmes the leadership team has time to look at the data, participate in meetings, read and reflect and reach their conclusions about the best course of action. Then they go into broadcast mode and wonder why employees are being so resistant. Neuroscience shows that people will be much more supportive of change if they have been able to come to their own insights about why this is the right course of action. We tend to like something much more when we have made an active choice to select it (Sharot *et al*, 2012). Choice and reaching our own insights are critical. Leaders need to recognize that they have been in a very privileged position, looking at the information and having time to make sense of it. They need to allow employees more time to do this. Not that there is always time – if there is a crisis, then leaders need to act immediately – but organizations do need to get cleverer at bringing employees with them and this is one means of achieving that end.

## Mindset workshops

One of the great benefits of Carol Dweck's *Mindset* is that it can be taught – to leaders, managers and all employees. Developing people's skills during times of uncertainty is a very positive thing to do as it builds confidence and demonstrates that the organization still wants to invest in its people. Feeling that we are learning, developing and becoming more skilled is rewarding to our brains. Research by Zhang *et al* (2018) suggests that learning can also help reduce stress as it builds our sense of being competent. A good starting point would be to teach people about neuroplasticity (see Chapter 2) – the brain's ability to restructure and carry on learning if we choose to apply ourselves. As we learn, the brain makes new connections and strengthens connections so that we become more skilled and faster at carrying out the action. Just the fact of knowing this makes a big difference to people, especially to older employees who might have told themselves that they have passed an age where they can learn new skills. A next stage might be to get people to reflect on a time when they have developed a new skill – it might have been hard initially but practice made a big difference. Researchers (Heslin *et al*, 2006) found that managers with a growth mindset are more likely to notice an improvement in their employees because they are aware that people can learn; managers with a fixed mindset on the other hand are less likely to notice improvements because they are stuck with their initial impressions. This mindset makes a big difference to managers' willingness to act as coaches – if they believe in a growth mindset, they believe in the value of coaching but if they believe that people's abilities are pretty much fixed then they believe there is no point in trying to develop people.

## Innovation and creativity

As previously discussed, great ideas don't happen at your desk. If we really want employees to come up with new ideas, we need to free them from being tied to their desks. Some organizations – Google with its sleep pods, Innocent with its grass and play areas – have created a work environment where people can rest and play and get their minds into a different state. If we want employees to be innovative, we have to trust them and not expect them to be at their desks. Let them go for a walk, let them take time out of the office.

Many organizations use 'brainstorming' as a way of coming up with new ideas. There are views on the benefits or not of this as a method of creating new ideas, but what neuroscience does teach us is that we should sleep on the ideas. Our brains make sense of our ideas overnight (Maquet *et al*, 2000) and we are far more likely to have that 'aha' moment after a good night's sleep.

# Summary of key points from this chapter

- For many organizations, people's brain power is all they have: leaders need to know how to keep people performing at their best especially during times of uncertainty.
- The threat/away and reward/toward states have a big impact on our brains and on our ability to think – the former is much stronger and lasts longer.
- The inverted U of performance shows that we need some stress but not too much: a fine balance is needed in terms of our brains feeling stimulated and challenged but not overwhelmed. Too much pressure closes down our prefrontal cortex.
- Dopamine is an important neurotransmitter that, in the right balance, helps us to learn, feel positive and focus.
- Having a growth mindset – believing in the brain's ability to restructure and learn (neuroplasticity) – makes a significant difference to our ability

to take on challenging tasks, to learn and to managers' ability to coach others.

- We have our best ideas when the brain is quiet and allowed to run free.

## What can we do?

- Set short-term goals that people can achieve. Achieving a goal is rewarding to the brain and sets people up well for the next task at hand.
- Remind people of past achievements.
- Enable people to win.
- Give praise and recognition – unexpected praise and recognition generates even more dopamine and so is even better.
- Perform small acts of kindness – the effect can be contagious.
- Provide some novelty to pique people's interest and to stop them from getting bored, but be aware that too much novelty can lead to a threat response.
- Have some fun – feeling happy and laughing can help productivity.
- Our brains crave certainty and information and like to reach their own insights.
- Set up Mindset workshops so that people can move from a fixed to a growth mindset.
- Innovation and creativity need a quiet mind – employees are unlikely to have their best ideas while sitting at their desks working on their laptops: encourage them to go for a walk or whatever works for them.

# Reflections and planning

We know that the threat state has a negative impact on our ability to work at our best, and a reward state leads to employees who can perform well, so:

- How do leaders and managers tend to work in your organization: do they tend to focus on what people haven't done right? Is there more belief in the stick than in the carrot? If so, what impact do you think this is having on employees?
- Take a look at the messages that are being sent out from leaders. Are they focused on threat or reward, or a combination of the two?

- If the culture of the organization does tend to focus on threat, what can you do to counterbalance this?
- Are leaders aware of the inverted U – that if you keep pushing people harder and harder, this can be counterproductive?

We know that reminding people of past achievements and enabling people to win puts them into a more positive frame of mind, so:

- When did you last talk about past achievements or, even better, ask employees to recall past achievements?
- Where can you break down tasks and set short-term goals so that people feel that the organization and they are making progress?
- How good are leaders and managers at helping people to be positive and feel confident in the midst of uncertainty?
- How can you set up situations where people can win or succeed so that you begin to create this positive cycle for them?

We know that our brains crave information and certainty, so:

- What can you do to improve information around the changes you are introducing?
- How well have you communicated the changes and the rationale for them?
- Do you know what issues are making employees anxious?
- Are employees clear about how and when the organization will communicate with them?

We know that a growth mindset has a positive impact on people's resilience and willingness to learn and grow, so:

- What's your mindset and what is that of the leaders and managers in your organization? We all have a fixed mindset about something: where does your organization have these limiting beliefs?
- How can you make people in your organization aware of the impacts of these mindsets?
- What can you do to share knowledge about the mindsets?
- How can you demonstrate the impact of the two mindsets? What examples of both are there in your organization that can bring the idea to life?
- Which leaders are good examples of people who learn from mistakes?

# References and further reading

Arnsten, A F T (2009) Stress signalling pathways that impair prefrontal cortex structure and function, *Nature Reviews Neuroscience*, **10** (6), pp 410–22

Aronson, J, Fried, C B and Good, C (2002) Reducing the effects of stereotype threat on African American college students by shaping theories of intelligence, *Journal of Experimental Psychology*, **38**, pp 113–25

Blackwell, L S, Trzesniewski, K H and Dweck, C S (2007) Implicit theories of intelligence predict achievement across an adolescent transition: a longitudinal study and an intervention, *Child Development*, **78** (1), pp 246–63

Bridges, W (1991) *Managing Transitions*, Perseus Books, New York

Chancellor, J, Margolis, S, Jacobs Bao, K and Lyubomirsky, S (2018) Everyday prosociality in the workplace: the reinforcing benefits of giving, getting, and glimpsing, *Emotion*, **18** (4), pp 507–17

Collins, J (2001) *Good to Great*, Random House, New York

Csikszentmihalyi, M (1990) *Flow: The Psychology of Optimal Experience*, Harper & Row, New York

Dweck, C (2007) *Mindset: The New Psychology of Success*, Random House, New York

Fink, A and Benedek, M (2014) EEG Alpha Power and Creative Ideation, *Neuroscience and Biobehavioral Reviews*, **44**, pp 111–23

Gruzelier, J (2009) A theory of alpha/theta neurofeedback, creative performance enhancement, long distance functional connectivity and psychological integration, *Cognitive Processing*, **10** (1), pp 101–09

Heslin, P A, Vandewalle, D and Latham, G P (2006) Keen to help? Managers' implicit person theories and their subsequent employee coaching, Personnel Psychology, **59**, pp 871–902

Hong, Y *et al* (1999) Implicit theories, attributions, and coping: a meaning system approach, *Journal of Personality and Social Psychology*, **77** (3), pp 588–99

Jordan, M [accessed 3 September 2015] 'Failure' ad for Nike [Online] https://www.youtube.com/watch?v=CgW48mBQJ14

Maquet, P *et al* (2000) Experience-dependent changes in cerebral activation during human REM sleep, *Nature Neuroscience*, **3** (8)

Moser, J S *et al* (2011) Mind your errors: evidence for a neural mechanism linking growth mind-set to adaptive post-error adjustments, *Psychological Science*, **22**, p 1484

Nakamura, J and Csikszentmihalyi, M (2001) 'Flow Theory and Research', in *Handbook of Positive Psychology*, eds C R Snyder, E Wright and S J Lopez, pp 195–206, Oxford University Press, New York

Nussbaum, A N and Dweck, C S (2008) Defensiveness versus remediation: self-theories and modes of self-esteem maintenance, *Personality and Social Psychology Bulletin*, **34** (5), pp 599–612

Olds, J and Milner, P (1954) Positive reinforcement produced by electrical stimulation of septal area and other regions of the rat brain, *Journal of Comparative and Physiological Psychology*, 47 (6), pp 419–27

Oswald, A J, Proto, E and Sgroi, D (2009) Happiness and Productivity IZA Discussion Paper 4645

Sahakian, B J (2014) What do experts think we should do to achieve brain health?, *Neuroscience and Biobehavioural Reviews*, 43, pp 240–58

Schultz, W, Dayan P and Montague, P R (1997) A neural substrate of prediction and reward, *Science*, 275 (5306), pp 1593–99

Sharot, T *et al* (2012) Is choice-induced preference change long lasting?, *Psychological Science*, 23 (10), pp 1123–29

Wise, R A (2004) Dopamine, learning and motivation, *Neuroscience*, 5 (6), pp 483–94

Yerkes, R M and Dodson, J D (1908) The relation of strength of stimulus to rapidity of habit-formation, *Journal of Comparative Neurology and Psychology*, 18 (5), pp 459–82

Zhang, C, Myers, C G and Mayer, D M (2018) To cope with stress, try learning something new, *Harvard Business Review*, 4 September

Answer to the word puzzle is *apple*.

# Our social brains

05

## The role of leaders and managers

**W**e have hugely underestimated our need for social connection at work. We have set up organizations and developed leaders based on misunderstandings and miscalculations that mean we have not yet created a work environment where employees can truly work at their best. Historically, organizations have valued an ability to plan, to devise strategy, to manage finances and to sell, and have placed less value on the ability to understand people and to nurture relationships. These latter areas have been termed 'soft skills' and have been seen as 'nice to have' but not necessarily a 'need to have' at work. When organizations decide to review their position or to embark on change, their first call tends to be to those consultancies that focus on strategy and structure. These consultancies might bring excellent thinking and great analytical skills but they tend to be less interested in the social and emotional aspects of change. The people aspects of change are considered but only as an after-thought and rather last minute. This is despite the fact that countless change programmes have failed because employees have chosen not to get behind them. I remember being called in to assist a global financial services company that had been working with a strategy consultancy for a number of months. By the time I was involved, the strategy had been developed with the support of the CEO but very few other employees had been involved, with the result that many leaders and managers felt excluded and alienated. Implementing the change was an uphill struggle. This brought home that when it comes to change, the leadership team might have developed an excellent plan but if they cannot communicate it to their employees, or do not have a strong enough connection with employees for employees to believe them and trust them, then the change won't happen. Neuroscience is beginning to provide us with a better understanding of our social brains and this knowledge means that we can design change programmes that work with our brains not despite them.

One of the reasons why the contents of this chapter matter is that there are many aspects of change that leaders and managers cannot control, but one thing we can all influence is the quality of our relationships with other people. We can choose whether to listen to others or not, we can choose whether to spend time with the team, we can choose whether we ask for opinions and we can choose whether to take an interest in a team member or not. During times of change, employees' need for connection with others becomes all the stronger. When we are uncertain about our future or are going through difficult times, we seek refuge in our relationships. This is well documented amongst war veterans – close bonds not only helped them through their traumatic experiences but have also been shown to help them in later years deal with the losses and suffering (Elder and Clipp, 1988).

However, there is a tendency when leaders are in the midst of change and uncertainty to pay less attention to relationships. Following significant announcements of change, leaders can disappear from view, sometimes literally leaving the work site. When asked about this, they give various reasons: they are busy trying to make the change happen; they are uncomfortable because people have lots of questions and they can't answer them; and/or they are worried about their own futures. However, this is the very time when connection with leaders matters more than ever to employees: leaders and managers have a crucial role in connecting, communicating, reducing stress levels and helping people to identify what they can still influence.

A recurring theme throughout this book is that connecting with people is not just 'nice to do': it has an impact on people's capacity to think and to work at their best. As outlined in Chapters 2 and 3, the threat state that change creates has a major impact on our ability to think and to stay calm. In this chapter, we will look at some of the work undertaken by the psychologist Roy Baumeister on the impact of social rejection on our IQ.

Another reason why understanding the social brain is going to become all the more relevant to organizations is that work is becoming more collaborative. Organizations rely heavily on teams, be they real, virtual, temporary or permanent. To deliver a product or great customer service, most employees have to rely on their colleagues to play their part too. The introduction of enterprise social media – online communication channels – means that employees are being given the tools to communicate and work with others across locations, regions and countries. Learning from each other and sharing ideas should lead to competitive advantage. But this will only work if employees are in a culture where collaboration is encouraged, where leaders understand the social brain and what our brains need in order to be willing to share and get involved. Even when the organization is not faced with significant change, it still needs people to feel they can talk to others and share ideas. One of the key differentiators between success and failure is the ability of organizations to innovate. You can have the best ideas in the world, but if you can't communicate with others, the idea won't happen.

One last point on why understanding our social brains matters: organizations struggle with the challenge of what skills and whom to promote. Many organizations used to promote on technical ability but then were faced with the issue of having leaders who were masters of their discipline but unable to get the best out of their teams. This is changing in some organizations with the view that the head of marketing, for example, does not need to be the best marketer but she does need to be the best person at leading others in the marketing team. The insights we are gaining from neuroscience underpin the need for leaders to have strong social abilities: emotional intelligence or EQ, as it is often called. The good news is that our ability to learn and grow (neuroplasticity) in this area carries on well into our late forties and fifties. Understanding the social brain is crucial to improving the lives of employees and to the success of organizations. The next section looks at some of the science behind this.

# The science

## *Survival*

As described in Chapter 2, much of what goes on in our brains is driven by the goal of survival, and the same is true of our need for social connection.

### The advantage of being part of a group

Recognition that people are motivated to form and maintain relationships is not new: as John Donne the English metaphysical poet recognized in the 17th century, 'No man is an island'. We feel connected to others and we feel a need to belong. Why have we evolved to want to belong to a group? What are the benefits to us? Once again, the answer to this question takes us back to our ancestors. For them, being part of a group had many advantages: a group of people could hunt for larger animals more easily than one person, together they could better avoid predators and fight off other tribes, provide offspring with more protection, and build shelter. Loners would have been at a severe disadvantage. So, in terms of survival, it makes sense that our brains should be motivated to connect to others. As we saw in

Chapter 2, evolutionary anthropologist Adrian Dunbar (1998) believes that the primary reason our brains have grown larger is so that we can live in larger groups, ie we developed a larger brain to be social.

## Mammals

There are other aspects of survival that come into play here. Human beings are mammals, and being a mammal means that without someone to look after us from the moment we are born we would not survive. Throughout our early years we are completely dependent on others to feed and protect us. From our first moment, we are constantly on the lookout as to whether someone cares about us. As babies, we don't just cry because we are hungry or cold: we also cry because of physical separation from our caregivers. Fortunately for babies, their cries also motivate their caregivers to want to look after them: their brains are wired to protect. This need for social connection carries on throughout life: it might be less overt than when we are very young, but the need to feel that someone cares about us and is interested in our welfare continues, including in the workplace.

## Why Maslow got it wrong

In his book *Social: Why our Brains are Wired to Connect* (2013), Matthew Lieberman, Professor and Social Neuroscience Lab Director at UCLA argues that Abraham Maslow got it wrong. Maslow, a 20th century American psychologist, proposed the now-famous hierarchy of needs in his 1943 paper *A Theory of Human Motivation*. The hierarchy is often depicted as a pyramid that sets out our needs and motivations in order of priority, from the base of the pyramid upwards.

Maslow's hierarchy of needs:

- Self-actualization – top of the pyramid
- Esteem
- Love/belonging
- Safety
- Physiological – bottom of the pyramid

The ones at the base of the pyramid, such as physiological needs (water, food, shelter, warmth), are essential, whereas those further up the hierarchy such as esteem and self-actualization are nice to have but we can live without them. Maslow suggested that we must fulfil our needs at the bottom of the pyramid before we can move up the hierarchy so our most pressing needs are physiological: we won't begin to concern ourselves with other needs until that one is met. Maslow said that our need for love, belonging and relationships does not emerge until hunger, safety and other key needs are satisfied. However, Lieberman challenges this: without love, care and social connection as babies we would not get our physiological needs, and nor would any other mammalian infants. We need other people to provide food and warmth. He therefore suggests that social needs should be moved to the bottom of the pyramid: being socially cared for and connected is crucial. We might think love and connection are 'nice to have' but without them we would not make it through our initial years. Our need for connection is paramount and this stays with us throughout life.

It could be argued that our social needs are even stronger today than those of our early ancestors. For our ancestors, meetings with lots of strangers would have been rare, but today we need to be expert at dealing with social situations, both in and out of work. We need to recognize other people and have an idea of what they might be thinking. To operate well in society, we have to be able to read others and understand what their intentions might be. These are some of the most complicated things our brains need to do. Some people find it difficult, such as people with autism and Asperger's. Some people lose this ability through damage or disease of the brain, such as dementia. Our modern day brains now have to be extremely good at navigating the social landscape.

Our need for good social skills is not confined to the fact that we have more people, and a greater variety of people, to deal with. The developed world means that few of us could get our basic needs met without dealing with others. Taking food as an example, hardly any of us are self-sufficient and get our food directly from the land any more: we need to go via a network of other people. We have to earn money and we then go to markets or shops to buy food; shops deal with wholesalers who need distributors; and so on. We are now very interdependent and to manage this interdependence we rely on our social skills.

## Social rejection hurts

Perhaps one of the most interesting and surprising findings to have emerged from neuroscience is that although we tend to think of social pain (bereavement, feeling rejected, being excluded by the group, being criticized by the boss) as being very different from physical pain, to the brain they are very similar. The brain processes physical and social pain using the same network. This network includes the dorsal anterior cingulate cortex (dACC) which deals with the distressing aspect of pain and both the dACC and the insula are active when we feel socially excluded; more activation of the dACC is associated with greater social pain. If you'd like to hear more about the network and how it deals with both physical and social pain, listen to Dr Naomi Eisenberger of UCLA (YouTube Yale Courses, Experts in Emotion, Social Pain and Pleasure, May 2013). She describes how she was analysing fMRI scans of people undergoing social rejection while sitting next to a colleague who was looking at scans of people experiencing physical pain. They realized the scans looked remarkably similar and in some cases, it was impossible to differentiate between the two. That the brain should process the two experiences via the same network is initially surprising but on reflection it makes sense: both are alerting us to threats to our well-being. Where the brain perceives there to be a threat to our survival and a need that must be dealt with, there is pain to make us pay attention and to act. That need might be emotional or physical. Social exclusion can be just as detrimental to our welfare as physical harm.

Subsequent experiments in the lab have reinforced this finding. Eisenberger, Lieberman and Williams (2003) asked participants to lie in an fMRI scanner and play a game called cyberball, a simple game of 'catch', where the participant tossed a virtual ball to two partners and they threw it back. Then, the other two (in fact, unbeknown to the participants, there are no other players – the virtual players were manipulated by the scientists) stopped throwing the ball to the participant: the participant is excluded from the game for ever. Eisenberger *et al* could see from the scans, that once the person realized they were being ignored, the pain network in their brain was activated. When asked after the game how hurt people felt by this exclusion, the more hurt people reported, the more activation the researchers could see in the dACC. Human beings are very sensitive to rejection: even something as small and unimportant as being excluded from a virtual game in a scanner can activate the pain network in the brain. Other research shows that within just 20 seconds of not getting the ball, we feel left out. We very quickly sense subtle rejection – both in the lab and at work. Even more intriguing is that just as we can take paracetamol or painkillers to reduce our physical pain, these painkillers have also been shown to be effective in reducing social pain (DeWall *et al*, 2010). Social pain is that real to the brain.

The fact that social pain is treated as physical pain by the brain is not only interesting in its own right but should cause leaders to reflect. Just as we would not expect someone with a twisted ankle to run to the next meeting, nor should we expect an employee who has just been given some negative news to run along to the next meeting and get on with their work. Organizations need to pay attention to this: our sensitivity to fairness is particularly heightened during change. Feeling that others are getting an unfair advantage over us, or that others are more highly thought of, activates the pain network in our brains. We think of social and physical pain as being very different, but actually they are not so dissimilar. Our brains have evolved to feel that threats to our social connections are experienced as pain. The pain we feel from loss of social connection means that we will always seek out connection and a sense of belonging, both in and outside the workplace.

Social pain endures longer: if I asked you to think about a time you had a headache, you might recall the experience; but if I asked you to think about a time someone insulted you or you had a nasty argument with someone you care about, you won't just recall the experience – the painful feelings will no doubt come flooding back as well. We tend to remember the emotional aspect of an event. We will come back to this in Chapter 8 on communication.

## *Ingroups and outgroups*

Think about the people you work with or have worked with in the past. There will be those with whom you have a natural affinity – things in common, shared interests: these are the people you probably choose to spend a bit more time with (ingroups). Then there are others who you perhaps feel are a bit different from you, and others again who you would prefer to spend as little time with as possible (outgroups). We all have ingroups and outgroups and we too are part of other people's ingroups and outgroups. As we will explore in more depth in Chapter 7 (decision-making and bias), our brains make decisions about people very quickly. Within seconds of meeting someone, we will quickly categorize people as ingroup or outgroup, friend or foe. Why does this matter? Because if people perceive they are not part of the ingroup and consequently feel socially excluded, it affects their ability to think. No doubt we have all had that experience: working with a boss with whom we don't quite get on, we know that for whatever reason they prefer some of our peers over us. It is not a comfortable feeling and sure enough it triggers a threat response in our brains.

### 'Us' versus 'them'

Belonging to an ingroup is extremely powerful. Such is our need to want to belong that even when we know that the group we are in has been put together in an arbitrary fashion, we still feel loyalty to that group and identify with group members very quickly. Henri Tajfel, Professor of Psychology at Bristol University, has done a great deal of research in this area, which we will look at in more depth in Chapter 7. His research shows how quickly we favour those who are in our group to outsiders (Tajfel, 1970). This can be both a positive and a negative. On the downside, this research shows how quickly and easily we can become biased, which is part of the focus on Chapter 7. On the positive side, it shows that being arbitrarily put into a group and given an identity can quickly lead to a bond between team members and has a powerful effect on behaviour. This kind of identity labelling can be used in a positive context in the workplace to improve productivity.

## Social skills improve team working and collective intelligence

Being part of a team is the norm for many of us at work – be it a permanent team, or one that comes together to focus on a specific task and then disbands. One of the questions that many organizations grapple with is what helps to create a high-performing team? Here again, the social brain influences how well teams work together. Chris Chabris, Associate Professor of Psychology and Co-Director of the Neuroscience Program, Union College, and colleagues have explored whether teams can have a 'collective intelligence' in a similar way to how we think of intelligence in individuals (Woolley et al, 2010). They found three key factors that correlated with collective intelligence:

1 Teams with members who contributed more equally in conversations had higher collective intelligence and performed better in group tasks than teams in which one or two people dominated.

2 Teams with stronger social sensitivity (ie better performance in a 'Reading the Mind in the Eyes' test) had higher collective intelligence than teams with weaker social sensitivity. 'Reading the Mind in the Eyes' measures how well people can read complex emotional states from images of faces when only the eyes are visible.

3 Teams with more women outperformed teams with more men. This was in part explained by the fact that the women in the tested sample, on average, were better at 'mindreading' (the ability to see things from others' perspectives) than men.

## Social rejection reduces IQ

Not only does social pain feel like physical pain to our brains, but it also has a major impact on our ability to think. Rejection can dramatically reduce a person's IQ and their ability to reason analytically, while increasing their aggression (Twenge *et al*, 2001), according to many research studies conducted by Roy Baumeister, Professor of Psychology at Florida University (Baumeister, Twenge and Nuss, 2002).

It is well established that rejected children tend to be more violent and aggressive, but Baumeister and colleagues found that randomly assigning people to experiences of rejection can lower their IQ scores and make them more aggressive. Baumeister *et al* (2002) conducted a range of experiments to test this. In one of the experiments, a group of strangers met, got to know each other and then were asked to list which two other people they would like to work with on a task. Each person was then told either that no one wanted to work with them or that everyone wanted to work with them. In another experiment, people were asked to take a personality test and were then given false feedback that told them that in the future they would either end up lonely or surrounded by family and friends. For those who felt rejected, IQ scores noticeably dropped. The researchers wanted to test out whether receiving bad personal news per se was the contributing factor, and so a third experiment was conducted in which participants took a test and were then given false feedback that they were accident-prone and ran a larger risk than others of seriously injuring themselves. These people performed better than those who felt rejected, leading to the conclusion that it is feeling rejected by others that causes the detrimental effect, not bad news per se.

In subsequent experiments, Baumeister *et al* (2005) found that people who felt rejected were more likely to give in to eating unhealthy sweet foods than the control group, and they gave up more quickly in unpleasant tasks, such as drinking a healthy but unpleasant tasting drink, or completing a difficult puzzle.

In summary, people who felt rejected:

- worked more slowly than others;
- achieved fewer correct responses in an IQ test;
- performed less well in memory tests;
- had greater problems with active reasoning;
- were less persistent;
- had less self-control;
- were more lethargic.

In short, feeling rejected causes a decrease in our ability to think intelligently. If we stop and pause here for a moment and think of our own experience at work, we have no doubt all had an unpleasant experience with a boss or colleague where we felt mistreated or unfairly excluded. We know that after an unpleasant experience, we didn't work at our best for the rest of the day, couldn't think straight, got distracted and perhaps reached for the cookie tin. Baumeister *et al*'s research strikes home because it resonates with our personal experience. The challenge for those leading change is to remember this and to plan how we can keep people feeling included and part of a team throughout the changes. The challenge for everyone who manages people is to think about who feels in their ingroup or outgroup and the impact of feeling in an outgroup on team members' brains.

## The reward network in the brain

So far in this chapter we have explored the vulnerable aspects of our social brains – our need to feel connected, respected and included. There is also a network in our brain that is activated by positive social experiences. The reward network consists of: the ventral tegmental area (VTA) – one of the areas of the brain that generates dopamine; the ventral striatum (VS); the ventromedial prefrontal cortex (VMPFC); and the amygdala (Amyg) – a part of the brain that is often associated with threat, but it also has a role to play in processing emotions. Just as the pain network is activated by both social and physical deprivations, so the reward network is activated by social and physical pleasures. Not only is it activated by eating chocolate, for example, but also by being treated fairly, cooperating and giving to charity. Humans cooperate with others because it makes us feel good and activates the reward network. At first, this might seem surprising: Lieberman and Eisenberger discuss this in their article, 'The Pains and Pleasures of Social Life' in *Science* (February 2009). Our need for survival might well drive the pleasure we feel in cooperating. Again, thinking back to our ancestors, the tribe had a better chance of surviving together than the loner. As Eisenberger and Lieberman suggest in their article, not cooperating might lead the tribe to oust a member which would reduce that person's chance of surviving, so it's in our interest that the brain makes cooperating feel rewarding.

# What can we do? Solutions and examples of what other leaders have done

It is clear that our brains are wired to connect with other people. We seek out others and we feel 'pain' when separated or rejected, and these feelings are heightened during change. Not only that, but feeling excluded has a negative impact on our ability to think, to reason, and on our tenacity. So, if we want people to perform at their best during change, we need people to feel connected at work and that they have a good working relationship with their manager. In one sense it is hard to be prescriptive as to what organizations need to do to ensure people's social brains are in a good place. This is where leaders and managers need to get to know their people and decide what will work best for each team member. Here are some examples of what leaders and organizations have done to try to achieve this:

## Our social brains – share the knowledge

The first step in any organization going through change (or in steady state) is to make sure leaders and managers are aware of the fundamental need we all have for a good working relationship with others. Being aware of this need, and the impact on our brains when we feel it is threatened in any way, is crucial. Many leaders and managers instinctively know that people will perform better when they have a good relationship with their boss, but they still feel that going out with the team for coffee, stopping for an informal chat, or taking the team out just for social reasons, are indulgent things to do and might be open to criticism from their peers.

I have worked with leaders in organizations where employee engagement was not the driving factor for leaders, but hitting targets and enabling employees to keep performing at their best in order for the team to reach their targets was of great interest. Just being aware that in some small way these leaders might be reducing people's ability to think and perform was enough to catch their interest and change their behaviour. Roy Baumeister's research can be extremely useful for getting leaders to sit up and pay attention. So, step one for any organization would be to make leaders aware of the social brain and what it needs to stay focused and positive.

## Who is in your ingroup?

One useful exercise we conduct in workshops is to ask leaders to think about who is in their ingroup. In part based on Dunbar's work, we ask the leaders to take a piece of paper and put themselves – 'Me' – in the centre and then draw circles around 'Me'. They then map on to the circles the people they are very close to (say four or five people in the inner circle), then those who are slightly less close in the next circle, and so on. We then ask them to reflect on the circles, and how satisfied they are with their connections.

Are there relationships they would like to make closer? Are there some team members who might be feeling excluded or part of the leader's outgroup? Remembering the impact of feeling rejected, what changes do they need to make?

One manager in a retail bank applied this thinking. As a result of learning about applied neuroscience, she changed her team meetings. Not only did she apply some of the points from the last chapter, such as starting team meetings with a reminder of some positive moments (activating dopamine and the reward network), but she also moved the location of some of the team meetings so that the team could start to network with other parts of the bank. As a result, employees got to know each other, and this led to a noticeable increase in the number of referrals of branch customers to other bank services.

Like many managers, one manager in an engineering firm recognized that he needed to widen his ingroup within the team and to rebalance time with team members. He decided to shift from, 'I'll see them when I need to' to, 'I'll see them because they are valued'. This shift, done in the right way, is subtly improving working relationships.

One leader in a not-for-profit organization, recognizing that a team member was being excluded by others, thought sharing the Eisenberger *et al* cyberball experiment with the team might be a useful way of raising people's awareness of the impact of exclusion. He reported that the fact that social pain is like real pain hit home, and knowing this has had a positive impact on those involved. Using neuroscience was a means to broach a sensitive subject and to get the team to be more empathetic and aware of just how sensitive we are to the slightest signs of rejection.

In another not-for-profit organization, a manager shared the insights on the social brain and on ingroups and outgroups with a team member who was having a problem with a key partner outside the team where there had been a difficult relationship. The team member worked on the relationship, applying some of the practical ideas in this chapter and in Chapter 7 on decision-making and bias, such as identifying shared goals and increasing face-to-face communication, and saw a significant improvement in the relationship.

## Connect to beneficiaries

In his book, *Give and Take* (2014), Adam Grant, Professor of Psychology at Wharton University, describes an experiment where he took the fund-raising team of a university and divided them into two. One group met a beneficiary of their fund-raising, just for five minutes, and the beneficiary told them about the impact of their work on his life. The other half did not meet a beneficiary. Grant found that fund-raising amongst the team that met the former student beneficiary increased by over 170 per cent, and not just in the following weeks but even into the following month, whereas the fund-raising of the control group remained unchanged. It is a simple story but it illustrates the impact of connecting employees to those who benefit from their work. There is a great deal of research into the human need to have a purpose at work and Grant's experiment is a simple but effective way of illustrating this.

When did employees in your organization last meet the customer (external or internal) or someone else who gains from their work? In most organizations, this is not hard to do and need not cost money, but Grant's research shows the enormous impact this can have. Two more interesting points to note from this research:

1 Before meeting the beneficiary in person, Grant gave the fund-raising team letters to read from beneficiaries. This helped but it had nowhere near the same impact as meeting the beneficiary in person.

2 Grant also reports that asking leaders and managers to deliver the same message about the impact of the fund-raising did not work. Employees need to meet the beneficiary in person and hear their personal stories. Second-hand storytelling is not enough.

So, connecting employees in person to those who have benefited from their work is vital for keeping people motivated. People want to believe and see that they are helping others: there is a huge untapped source of motivation here for organizations and employees. Therefore, bringing customers into the workplace or employee conferences, or enabling all employees to get out and meet customers, is an investment well worth making.

## 'Together'

Neuroscience demonstrates that the old saying, 'Sticks and stones may break my bones, but words will never hurt me' is wrong. We are acutely attuned to whether we are accepted or rejected: leaders and managers need to choose carefully the words they use during change.

Research by Priyanka Carr and Greg Walton of Stanford University in 2014 has revealed a way to put our brains into a positive social place, simply by the use of the word 'together'. In their research, people worked initially together and then moved on to work on difficult tasks separately.

One group was told that they were working on the task 'together' with others, even though they were in a separate room; the members of the other group were just left alone. The participants who had been told that they were working on the problem 'together' worked 48 per cent longer and solved more problems correctly than the members of the other group. They also enjoyed the tasks more – an important point in terms of employee engagement. The very word 'together' seems to activate the reward centre of the brain, making us feel we belong. The researchers commented that it does not take an enormous effort to create this feeling of belonging. We just need subtle cues that we are part of a team. People leading change programmes should make good use of this word (and mean it, not just using empty rhetoric) and other words that make people feel that they belong. Paying attention to making people feel part of a team boosts effort and performance. We will discuss the impact of language further in Chapter 8 on communication.

## Activating the reward network

As Lieberman's and Eisenberger's article in *Science* (2009) identified, the reward network in our brains is activated by various triggers in addition to physical pleasures.

### Being treated fairly

Believing that we are being treated fairly helps to put our brains into a more positive place. This need to be treated fairly increases during change: if things are going to be different, we feel that we want to have as good a chance as colleagues at getting the role that we want, or the pay that we deserve. So, those leading change need to check constantly whether the processes they are putting into place are fair. Developing a set of principles about how the change process will be handled and then using these principles as a constant reminder and checklist is useful both for those orchestrating the changes and employees. It helps to build transparency and certainty. Fairness needs to be one of those principles, and in every meeting where decisions are being made decision-makers should check whether both the process and the outcome will be fair.

### Giving to charity

There's a phrase, 'it is better to give than to receive', and science shows our brains find giving, rather than receiving, rewarding, so this phrase might be true in more ways than one. Elizabeth Dunn *et al* (2008) conducted research showing that spending on others rather than oneself led to greater happiness. Many organizations have charities that they and employees contribute to. In one organization, employees decided to get more involved with charity work as their organization was going through significant change. The employees said that doing something for people who were less fortunate than them made them feel better during a difficult time. It also gave them something positive and constructive to do during a change programme where they felt they had little control (sites were being closed down and jobs were being lost). The fact that it had been their decision to raise money for charity made the work all the more positive. As we will see in Chapter 8, our brains respond positively to autonomy and choice.

## Recognition and having a good reputation

People prefer achievements that are validated, recognized and valued by others over solitary achievements. Just watch children in a swimming pool, shouting to their parents 'Look at me'. Children are open about their need for validation. As adults, most (but by no means all) people are slightly more inhibited about their need for validation. Feeling recognized and respected is rewarding to our brains and this in turn puts our brains into a better place. Most of us don't call over to the boss to say, 'look at what a great job I am doing here', but it feels good when someone recognizes our work (and it feels even better when the recognition is unexpected, as we saw in the last chapter). Who is doing a good job in your organization but perhaps not getting the recognition they deserve? Recognition is not just 'nice to do' – it triggers the reward network in the brain, which in turn helps people to perform better.

## *Make time for people*

Having learned more about the social brain, many leaders recognize that they need to spend more time with their team members. This can be difficult for leaders for many reasons – lack of time, geographically-scattered employees, personal shyness. But neuroscience shows us that people have a strong need for social connection, and without it they will constantly be in a threat state and unable to focus at their best. Leaders need to focus on their people but this needs to be done sensitively. One leader said he went straight from a workshop which a colleague and I ran and booked in a one-to-one meeting with each member of his team. Not knowing what the purpose of these meetings was, team members went into a threat state, assuming he must be about to impart bad news!

It should go without saying but it needs to be said nevertheless: people want to feel that someone is listening and this is especially true when going through change. To this day, I can hear the heartfelt words of an employee in a large government department who had been relocated and moved to another team – 'I just wanted someone to listen'. She was not looking for answers – she just wanted to feel heard. She recognized that that would have been a step forward in helping her to deal with the changes.

## Helping people to let go

Social connections matter to us and having to let go of connections is one of the hardest things we have to do, bereavement being the most extreme. Change often means that teams have to disband and team members have to join new groups. In a hurry to get on with the new, we often don't allow employees enough time to let go of those old social bonds; and yet if we don't let people acknowledge the loss of the team, it can be difficult for people to start afresh. In one organization I have worked with, a team leader commented that she had two new team members join her from an office that had closed down. These two were the lucky ones – they had kept their jobs when their former team members had lost theirs – so why were they withdrawn and negative? Conversations revealed that psychologically they had not yet let go of the past.

Before teams disband, it is very useful to gather the team together in some way. Often the team members themselves are the best people to decide what they want to do and how (and as we will see later, giving them choice is very important). This is part of the ritual of helping people to let go of those social connections.

## Teams – belonging and shared goals

The shared thread amongst many of these activities is to make people feel that they belong. Belonging is a powerful motivator to our brains. Giving the team a sense of identity and shared goals (this is key in uniting team members – focusing on what we have in common), with regular feedback on how they are performing, can make a big difference.

A few years ago I was working on a merger in a pharmaceutical company in the United States. The Communications Director had two sets of consultants working on projects: often he would give us the same brief to see which group of consultants came up with the best idea. He believed a bit of competition would bring out the best in us. In reality, the competition led to us being protective of ideas and not wanting to share with the other consultants. We suggested to the client that he changed the way of working: that the two sets of consultants should be set a common goal and should work together. This subtle shift made a huge difference: relationships between individual consultants greatly improved. There was more energy and focus: together we were focusing on the client's challenges rather than being competitive. The threat state was removed and we worked much better.

## Summary of key points from this chapter

- We have hugely underestimated human beings' need to feel that they belong and are connected to others. This need increases during times of uncertainty.

- There are many aspects of change that leaders and managers cannot control, but we can all influence the quality of our relationships with others.

- Work is becoming more collaborative and so our ability to navigate the social landscape is of increasing importance.

- In terms of survival, it is advantageous for humans to be part of a group. Loners were less likely to survive on the savannah. Arguably, our need to be able to understand others is even greater today, especially in the developed world, as we have become very dependent on others for our most basic needs of food, water and shelter.

- As mammals, we would not survive our earliest days if we did not have someone to protect us and provide for us. It is therefore wired into us to be constantly on the lookout as to whether someone is looking after us and interested in us. This continues throughout life, including in the workplace.

- Our brains process social pain using the same system as they do for physical pain. To the brain, social rejection is pain. The pain system is an alarm that tells us that in some way our well-being – physical or social – is threatened and we need to act.

- We are highly sensitive to social rejection – even feeling excluded from a game of cyberball activates the pain network.

- Research into team working shows that the most successful teams are those that have people who are willing to take turns in talking and who are good at seeing things from others' perspectives.

- We all form ingroups and outgroups but feeling part of our manager's outgroup will affect our performance. Being part of an ingroup, on the other hand, with a sense of belonging and clear, shared targets, can be hugely motivating.

- Experiments conducted by Roy Baumeister demonstrate the impact of rejection and exclusion on our ability to perform: social rejection reduces IQ.

- Just as the pain network in the brain deals with social and physical pain, so the reward network is activated by physical and social pleasures. These social pleasures include cooperating, having a good reputation, and giving to charity.

## What can we do?

- Share this knowledge about the social brain with leaders and managers. Leaders who understand the brain's fundamental need for social connection will be better placed to get the best out of people.

- Leaders and managers need to think about and map out who is in their ingroup and what they need to do about their relationships at work. This is an area they can easily influence.

- Connect employees to those who benefit from their work. Human beings like to help others and understanding the positive impact we have on others gives us purpose and is rewarding to the brain. This works best when employees meet the beneficiary in person: written statements and second-hand reporting via leaders and managers does not have the same impact on motivation.

- Making people feel part of a team is important. The very use of the word 'together' can have a positive psychological impact on people.

- Activating the reward system does not have to cost money but can have a positive impact on people's ability to perform.

- Given the importance of social connection, make time to be with people. Just feeling listened to can make a big difference.

- Dismantling teams and losing former team members can be difficult because of the social loss. We need to help people let go of the old team so that they can bond with new team members.

- Belonging to a team and having shared goals is rewarding to our brains.

# Reflections and planning

We know that our brains are wired to be social, and social rejection has an impact on our IQ:

- What gets valued in your organization: do leaders and managers feel recognized for their ability to lead people?
- Are leaders aware of the impact of social rejection on people's speed, accuracy, ability to reason, persistence and self-control? If not, how are you going to make them aware of this? What examples of this do you have in your organization?

We know that people form ingroups and outgroups:

- Are leaders in your organization aware of ingroups and outgroups?
- If so, have they analysed who falls into which group?
- What can you do to help them increase their ingroups?
- When did they last speak to each team member?
- What can they do to remove any barriers to people feeling part of their ingroup? What steps can they take to connect with each one?
- Do leaders know what matters to each member of the team, what motivates (and what frustrates) them?

We know that connecting people to beneficiaries is very motivating and has a positive impact on performance:

- Are people aware of the positive impact their work has on the lives of others?
- When did they last meet a beneficiary?
- How and when can you set up these meetings?

We know that the language we use matters:

- How are you communicating about the changes?
- Take a look at the written language you are using – does it feel inclusive?
- What about how your leaders speak? Does their language make people feel they are in it together?

We know that activating the reward network in the brain has a positive impact on our ability to think and collaborate:

- What are you doing to activate the reward network?
- Would employees say that the approach to change is fair – both the process and the outcome?
- Cooperating and helping others is rewarding to our brains: where are the opportunities to enable employees to do this during change?

We know that people's desire for social connection also means that people need to feel heard:

- What opportunities are there for employees to talk about how they feel?
- How good are leaders at listening? How comfortable do they feel listening?

We know that belonging to a group, having a shared identity and shared goals are hugely motivating:

- Which teams do people in your organization feel they belong to?
- Is there a shared identity?
- Do goals unite people or divide them?
- How frequently do people get feedback on those shared goals?

We know that social loss can be painful:

- Which groups will need to disband because of the changes you are introducing?
- What are you doing to help them deal with this loss?
- Are there opportunities here to involve employees in deciding how they let go of the former team?

# References and further reading

Baumeister, R F, Twenge, J M and Nuss, C (2002) Effects of social exclusion on cognitive processes: anticipated aloneness reduces intelligent thought, *Journal of Personality and Social Psychology*, **83** (4), pp 817–27

Baumeister, R F *et al* (2005) Social exclusion impairs self-regulation, *Journal of Personal and Social Psychology*, **88** (4), pp 589–604

Carr, P B and Walton, G M (2014) Cues of working together fuel intrinsic motivation, *Journal of Experimental Social Psychology*, **53**, pp 169–84

DeWall, C N *et al* (2010) Acetaminophen reduces social pain: behavioral and neural evidence, *Psychological Science*, **21** (7), pp 931–37

Dunbar, R I M (1998) The social brain hypothesis, *Evolutionary Anthropology*, **6**, pp 178–90

Dunn, E W, Aknin, L B and Norton, M I (2008) Spending money on others promotes happiness, *Science*, **319** (1687)

Eisenberger, N (May 2013) https://www.youtube.com/watch?v=gqlJy4nBQAc

Eisenberger, N I, Lieberman, M D and Williams, K D (2003) Does rejection hurt? An fMRI study of social exclusion, *Science*, **302** (5643), pp 290–92

Elder, G H Jr and Clipp, E C (1988) Wartime losses and social bonding: influences across 40 years in men's lives, *Psychiatry*, **51** (2), pp 177–98

Grant, A (2014) *Give and Take*, Orion Publishing, London

Lieberman, M D (2013) *Social: Why our Brains are Wired to Connect*, Oxford University Press, Oxford

Lieberman, M D and Eisenberger, N (2009) The pains and pleasures of social life, *Science*, **323** (5916), pp 890–91

Maslow, A H (1943) A theory of human motivation, *Psychological Review*, **50** (4), pp 370–96

Tajfel, H (1970) Experiments in intergroup discrimination, *Scientific American*, **223**, pp 96–102

Twenge, J M *et al* (2001) If you can't join them, beat them: effects of social exclusion on aggressive behaviour, *Journal of Personality and Social Psychology*, **81** (6), pp 1058–69

Woolley, A W *et al* (2010) Evidence for a collective intelligence factor in the performance of human groups, *Science*, **330** (6004), pp 686–88

# Managing emotions during change

## 06

**W**ho hasn't had an emotional outburst that they have later come to regret? We have all had moments when we have responded in a way that later we wish we hadn't: a colleague says something that normally we would take in our stride, but today we found the comment particularly irritating and we say so; we fire off an angry e-mail and as soon as we have pressed 'send', we know that it was a mistake; we have resisted temptation and eaten healthily all day and then we are feeling a bit tired and can't resist a chocolate muffin. Staying in control of our emotions is hard. We're not very good at understanding what causes us to be in control some of the time and then not at others: the subconscious brain remains a mystery to us. Fortunately neuroscience is beginning to throw light on what causes our emotions and what helps us to stay in control of them. Emotions have a purpose. To thrive at work and particularly during periods of change, we need to benefit from what is good about having them and we need to learn to avoid the less helpful aspects. Becoming more aware of our emotions and our responses enables us not just to react but to choose how we respond. We can make more informed choices about our behaviour. Leaders have a particular responsibility to do this, as we will explore in this chapter.

## Emotions – what are they for?

Emotions certainly have their uses. An advertisement for a children's charity features a picture of a small girl, probably about four years old, looking sadly at the camera. Why did the charity choose to focus on just one child? Logic might suggest that charities should tell us about the numbers of mis-treated children, the statistics, the number of hospital admissions or even deaths, and why they need our money to stop cruelty to children. But charities and their advertising agencies know that it is not logic but emotion that will lead us to give money. We might think of emotions and decisions as being quite separate but they are very much entwined: emotions are central to our decision-making. Without emotions, we find it very hard to make decisions. The University of Southern California neuroscientist, Antonio Damasio (2006), describes working with a patient who had damage to the part of the prefrontal cortex (PFC) which processes emotions. The patient found it very difficult to decide which restaurant he wanted to go to: his intelligence was intact and he could weigh up the pros and cons of different restaurants, but being unable to access his emotions, he did not have that final 'nudge' of emotion that told him which restaurant he'd prefer to go to. Damasio says that the same patient found it difficult to choose when he'd like his next appointment, spending tens of minutes analysing which of two dates might be best. Emotions give us an essential prompt to help us to make decisions (Damasio, 2006). Phineas Gage, the unfortunate railroad worker who damaged his frontal lobes (mentioned in Chapter 1), became emotionally volatile, and although he could make decisions they were no longer reliable or sensible due to the frontal lobe damage.

By guiding us as to what feels right and wrong, emotions are a crucial part of our decision-making. They help shape and change our behaviour. Emotions also play a central role in enhancing our memories. Our brains have developed such that if an event is particularly emotional, be that positive or negative, we are more likely to remember it (except in some extreme cases of trauma). The brain does this so that the next time we face a similar situation the memory will guide our behaviour.

Emotions are also crucial in helping us to navigate through the social world as we explored in the last chapter. They help us to understand where we fit and how to respond appropriately to other people.

Emotions have an essential role to play in our lives but they are not always useful. They are not helpful, for example, when they are the wrong type of emotion, appear at the wrong time, or are of the wrong intensity. They are not helpful when they lead you to shout at a team member, be dismissive of someone else's anxieties, or panic when a decision needs to be made swiftly and calmly. Sometimes we want to tone down the emotion, eg anger, feeling very sad or feeling mean towards someone, and at other times we want to increase the emotion – such as celebrating the team's achievement, sharing good news about winning new business, or congratulating a team member who has just passed an exam.

# Why does this chapter matter?

The contents of this chapter matter for many reasons:

1 As we know from Chapter 3, change means that we cannot predict, and therefore our brains are more likely to be in a threat state. This threat state means we are more vulnerable to anxiety, anger, and feeling insecure and in competition with our colleagues. Managing our emotions at work is always important but it is an even greater challenge during change when our sensitivity to threat and negative emotions is heightened.

2 There are plenty of causes of stress and emotion at work these days. For many of us, just getting to work can be difficult enough – busy roads, crowded trains and unreliable public transport systems mean that our limbic system is activated before we even get to work and our bodies are full of adrenalin and cortisol – stress hormones. Then, when we arrive, we have colleagues to deal with, whether we like them or not. People who work from home might not have to battle their way to work but they have to manage relationships with people whom they rarely or never see; as we know from the last chapter on our need for social connection, this can be difficult. It can be much harder to resolve issues with people we don't see and hardly know. Conflicting demands at work mean that we constantly have to prioritize and make judgements about what to do next – prioritizing and making decisions depletes our brain's resources, as we will see later in this chapter. Technology has many advantages but it also means that we are always 'on' – people can reach us via devices at any time and we don't have the same ability to relax. In addition to all this, people have to balance the needs of personal lives and the demands of work. This all adds up to a lot of demands and a lot of stress on 21st century employees.

3 There's a long-term detrimental impact of not managing emotions – of being constantly stressed, angry or suppressing them (we'll explore suppression in the Science section in this chapter).

4 Emotions are contagious, especially those of leaders. When leaders fail to control or regulate their emotions, employees pick up the emotion. Those leading change owe it to themselves and to others to stay calm. It's part of a leader's responsibility not only to manage their own emotions but also to help others manage theirs.

5 But it's not all bad news. The good news is that we can have more control over our emotional responses than most of us have recognized. We have

automatic reactions that arise from our subconscious, but we can become aware of them and modify them if we learn how to do this and practise it. Emotional responses are interpretations – we need to learn to take control of our interpretive processes.

This chapter will share techniques on how to stay calm in a constructive way. Suppressing negative feelings is not helpful to us or others, but there are other useful techniques we can learn, eg practising mindfulness (I will explain what this is, and give some tips on how to get started later in this chapter), labelling emotions, and reframing the situation to name just a few. Moreover, learning how to quieten our emotions gives us choice over how to respond and enables us to make calm, informed decisions. Being able to choose how to respond gives us power and is the mark of a good leader.

# The science

Most of us have felt happiness, surprise, fear, sadness, disgust and anger at some point in our lives – these are the six basic emotions as identified by Ekman *et al* (1969). Emotions usually arise quite suddenly due to thoughts, activities and interactions with other people. They are different from moods in that the latter last longer. Emotions are physiological responses to stimuli, helping us to move towards rewards and avoid dangers, ie their role is to prompt us to behave in a certain way that is in our interest. We are constantly generating emotions but we are largely unaware of them. In this section on the neuroscience behind emotions, we will look at some of the brain areas that deal with emotions, the impact of emotion and stress on us physically and mentally, why emotions are contagious, the insights neuroscience brings to constructive ways of managing our emotions, and techniques best to be avoided.

## *Critical systems that manage emotions*

Emotions activate widespread areas of the brain but they are largely associated with the limbic area of the brain. This is the part of the brain that has a role in recognizing emotions and influencing how we respond to them physically and mentally. One of the key areas is the amygdala. Tiny and almond-shaped, it processes stimuli with an emotional impact. It is particularly associated with fear. It often operates under the radar of conscious control and sends signals to other areas of the brain and body for an immediate bodily response to get us ready for fight or flight, as we saw in Chapter 3.

Another part of the limbic system is the hippocampus, which is associated with memory. Studies show that emotional arousal is linked to episodic memory – episodic memory is when we remember a specific event. Events that result in a strong emotional response, as opposed to little or no emotional arousal, are more likely to be important for our survival and well-being, so our brains use emotions to ensure we don't forget. To this day, I can remember a new business meeting that took place over 10 years ago where the potential client spent the whole time looking at his computer screen, sending messages on a portable device, not making eye contact, and generally making it clear that he was not interested in a word I had to say. I felt annoyed and angry but it also means that I have made a mental note – if someone behaves like that again, walk away.

The hypothalamus has a coordinating role in the limbic system and plays a key part in triggering the stress response. The limbic system is closely connected to the brainstem which regulates the body's systems (breathing, heart rate, etc). Some of the parts of the brain that play a key role in how we manage our emotions are shown in Figure 6.1.

**Figure 6.1**   Critical systems in managing emotions

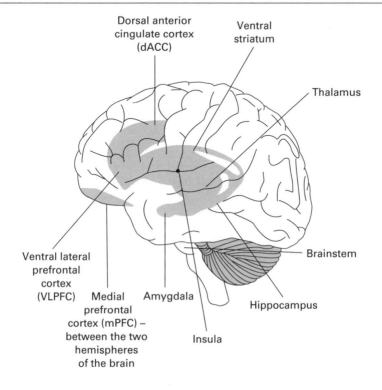

There are two other parts of the brain that are important to mention here as they will appear later in this chapter. One is the insula which is located in a fold beneath the frontal and temporal lobes. It picks up on activity in internal organs, especially negative ones. The second part of the brain that we will refer to again later in this chapter is the ventrolateral prefrontal cortex (VLPFC) which is one of the parts of the brain that acts like a braking system. The VLPFC appears to modulate the amygdala's response to threat (Monk *et al*, 2008). As its name suggests, it is at the front of our brains, behind our foreheads. The VLPFC is one of the parts of the brain that is important in emotional control – it stops us from blurting out to our idiotic colleague what we really think of him. However, its capacity is limited and it is quickly depleted – an important point that we will come back to.

## *Psychological stress*

When scientists want to study high levels of stress hormones, they simulate a job interview in which an applicant receives intense face-to-face criticism or they ask people to speak publicly in front of a critical audience. When we receive criticism from a person whose opinion matters to us, our stress hormones, adrenalin and cortisol, soar and our heart rate rapidly increases by 30–40 beats per minute. This is fight or flight kicking in – they can be useful in the short term but not so good in the long term. As we saw in Chapter 3, adrenalin and cortisol help us to respond to a short-term threat (Sapolsky, 1998). They help prepare the body to survive. When we were on the savannah, we just needed a short boost of adrenalin and cortisol to escape the predator. It's not meant to stay in our system for hours, days, months. However, in our current corporate world, the threat never goes away and so we have stress hormones constantly in our systems. In the long term, this is damaging to us physically and mentally.

In Chapter 4 we saw that some stress is useful to us – it gives the impetus to get on and get things done. It's also true that overcoming some stressful situations is beneficial because we learn tactics to deal with the challenges: we adapt and this helps to build resilience. But even in the short term, stress can prevent us from thinking clearly: our brains become so focused on dealing with the perceived threat that we are distracted, overly anxious and emotional and we start to look at the world much more negatively. When we are under stress, we tend to remember negative events and thoughts. We also need to be aware that unconscious fear also activates the amygdala: so we might not be aware of a fear or stress but our thinking will still be negatively affected. Stress depletes our self-control and our ability to control our emotions.

## Longer-term impact of stress

It is well established that stress has an impact on most people in the longer term. Although a few people do seem to have greater resilience, studies have established a link between psychological stress and diseases such as cardio-vascular disease and depression (Cohen *et al*, 2007), impaired immunity, atherosclerosis, obesity and bone demineralization. Prolonged stress over time leads to wear and tear on the body – this is sometimes referred to as 'allostatic load', a term created by Bruce McEwen, neuroscientist at Rockefeller University. Short-term stress can boost our immunity (Herbert and Cohen, 1993) but long-term stress has a negative impact on our immune system. Stress has become one of the main causes of prolonged absenteeism from work and common mental health conditions are reported to be increasing amongst employees (CIPD, 2018).

But it's not just the impact on us in terms of physical health that we need to be concerned about: stress hormones are toxic to neurons over the longer term, particularly in the areas that are key for high performance at work. The hippocampus (learning and memory) was one of the first areas in the brain to be identified as being vulnerable to stress hormones (McEwen *et al*, 1968). The hippocampus has lots of receptors for adrenalin and cortisol, and cortisol damages and destroys brain cells: too much stress and there is a danger that our ability to make and recall memories will be impaired. The prefrontal cortex, the executive centre of the brain which is so important for thinking, decision-making and self-control is also a target for stress hormones. People who are constantly stressed can develop an oversensitive amygdala (fear and anxiety) which is then too easily triggered by less and less pro-vocation. Research conducted by Ganzel *et al* (2008) showed that otherwise healthy people living close to (within 1.5 miles of) the World Trade Center buildings on 11 September 2001 had a reduction in grey matter in the hippocampus, amygdala and medial PFC three years after the attack.

## Boredom is a stressor

When we think about stress we tend to think about demanding environ-ments, tight deadlines, too many pressures, and a shortage of resources and means to deliver what is expected from us. But it's important to note that boredom can also be a cause of stress: dull, predictable, repetitive work is stressful. One of the challenges our brains face is keeping our attention focused. Our brains like a bit of novelty because it keeps them intrigued. When nothing is varied and things are monotonous, the brain struggles to stay focused and this is another source of stress.

## Continual change

One of the frequent observations of people working in organizations is that it is not so much the major change programme that stresses them – it's the layers upon layers of insidious change. One big change might be manageable, but lots of constant, unpredictable, seemingly small changes are what cause the real strain. Constant pressure mixed with uncertainty is very stressful. The combination of the two means that our prefrontal cortex, which is responsible for 'executive function', is overloaded. Our brains are 'limited capacity machines' and as we overload them the limbic system interprets this as a threat: the machine is under strain and can't function properly. As we saw in Chapter 3, this takes us into a negative cycle where performance declines. Edward Hallowell in his 2005 *Harvard Business Review* article describes this as 'attention deficit trait' (ADT): modern office life is turning steady executives into frenzied underachievers, he says. He points out that the symptoms of ADT come over us gradually as more and more pressures build, but we feel that we need to look and be resilient and to deal with them. However, we are in a constant low level of panic and guilt. Mentally and physically, this begins to take its toll.

## *Emotions are contagious*

We pick up others' emotions. Several studies have shown that team leaders in particular play an important role in influencing the emotional climate of the group. These studies include research conducted by Thomas Sy, Associate Professor at University of Michigan. He and his colleagues have demonstrated that groups 'catch' the mood of their leaders (Sy, Côté and Saavedra, 2005) and, once caught, groups' emotions have an impact on outcomes, with positive emotions giving the team a better chance of performing well.

How do we pick up these emotions? There are various ways in which we do this. The more obvious ones are through visual cues. Just seeing a photograph of an angry or fearful face leads to increased activity in the amygdala. The amygdala is activated even when the photographs of emotional people are shown so quickly that the image does not register in the conscious mind (Morris *et al*, 1998). Consciously or subconsciously, we observe people's behaviour, the movements and gestures that people make, their posture: we feel nervous because we see others are nervous. Other means of spreading emotions include auditory cues: the language people use and the tone of voice. Positive emotions can be contagious too: feeling happy and wanting to laugh because we see and hear others laughing (hence canned laughter on television and radio programmes). A key point to note is that our behaviour might be voluntary and conscious or it might be involuntary. We might be aware of the emotions we convey or we might not.

Mirror neurons might be another means by which we tap into others' emotions. Mirror neurons were only identified in the early 1990s when Italian neuroscientist Giacomo Rizzolatti and his team realized that some neurons were not only firing when a monkey moved to grab food but also fired when the monkey saw the researcher reach out for food (Rizzolatti and Craighero, 2004). Mirror neurons fire when we observe others making a movement but intriguingly only when there is some intent or purpose to that movement. The role of mirror neurons is still being debated. It might be that their role is to help us understand the actions and behaviours of others so that we are better able to live, work and survive with them. They might have a role in learning through enabling us to imitate the actions of others. There is also a theory (Gallese and Goldman, 1998) that we might pick up on others' emotions through mirror neurons.

As we will see later in this chapter, suppression of emotion can be costly for both the person who is suppressing the emotion and the person with whom they are interacting, as it can cause stress in both (Butler *et al*, 2003). There are many ways in which we deliberately or inadvertently share our emotions with others. Emotions, as we know, have an impact on our ability to perform at work and so leaders, managers, indeed all of us, need to become more attuned to our emotions and learn how to manage them so that they help, not hinder, not only us but our colleagues too.

## 'Best in show' – hormones and competition

Emotions are contagious – even to our pets. Research by Jones and Josephs (2006), Mehta *et al* (2008) and Sherman *et al* (2016) looked at the impact of cortisol on behaviour. The studies looked at how people and their dogs respond to qualifying or not for a second round of a 'dog agility' competition, where the owner has trained the dog to navigate obstacles. For those who did not qualify, it had a negative impact on their status and was a cause of stress: cortisol rises. This was particularly true of men with high basal testosterone. In these studies, men and women tended to respond differently to winning and losing. Men who won tended to pet their dogs more than women who won. Men who lost tended to pet their dogs less than losing women. In fact, the higher the level of the woman's cortisol, the more time she spent petting her dog. This is the 'tend and befriend' stress response. Jones' research showed that the dog's cortisol level rose in response to how their owner behaved – the 'stress contagion effect': cortisol rose most in the dogs of losing men whose testosterone dropped the most.

These research studies touch on all sorts of interesting points. In this research, women and men tended to behave differently after winning and losing: more focus on shared celebration amongst men, more focus on shared consolation amongst women. As in organizations, there is a hierarchy between the dog and the owner (the dog does not choose to enter the competition, the owner does); the dog's cortisol level is affected by the cortisol response and behaviour of the owner.

Physiological stress contagion is not necessarily bad: sometimes leaders need to put people 'on alert' when the organization is facing a genuine threat and people need to respond. Stress hormones mobilize energy and can help focus our attention; they make us more vigilant. The problem comes when leaders are constantly triggering cortisol – employees become worn down as a consequence and performance suffers.

## *Self-control*

When we think of self-control, we tend to think of it in relation to things such as not overeating, focusing on our work, and taking enough exercise. However, these are probably not the reason why we have developed self-control. The theory is – and this fits with much of what we have explored in terms of the social brain in the last chapter – we developed the ability to control ourselves so that we can get on with the rest of the group. For ancestors, self-control would have helped them to get on with the tribe – and the same is true of us today. Much self-control operates unconsciously – we are not aware of our brains stopping us from taking all the food at the table or stopping us from blurting out insults and nasty comments.

### Our self-control faces a range of challenges

Pause a moment and think about your day so far: what kinds of self-control have you had to exercise already? Have you had to deal with an annoying colleague? Have you had to get on with some rather dull work rather than going outside and taking a walk in the sunshine or surfing the internet? Self-control to deal with different things – people, eating, exercising, focusing on work – might feel quite different but, according to social psychologist Roy Baumeister, we have one 'tank' of self-control, and using it not to eat chocolate reduces our ability to keep going at a difficult piece of work.

## One theory: self-control is a limited resource

One theory currently being debated (Friese *et al*, 2018) is that our self-control equipment is fragile and quickly depleted. In their book *Willpower* (2011), Baumeister and Tierney describe The Radish Experiment – an experiment conducted by Baumeister amongst students. Before the experiment, students were asked to come to the laboratory hungry. In the lab were warm freshly-baked cookies, chocolate and radishes. Some students were offered cookies and chocolate to eat, the less fortunate ones were offered radishes. They were left alone to eat and those who were assigned radishes were also left with cookies in the room to test their self-control even further. These students were alone but the researchers could observe them through a small window: they saw the students look at the cookies, even pick them up and smell them, but they refrained from eating them. Clearly their self-control had been put to the test. The students were then taken to another room where they were given geometry puzzles. The students were told that the researchers were looking at cleverness but actually the puzzles were impossible to solve. The test was in fact about perseverance. The group that had eaten cookies and chocolate gave up on average after about 20 minutes, as did a control group who had been given nothing to eat, but the group that had eaten radishes and resisted the cookies gave up on average after just 8 minutes – a huge difference in terms of lab tests. Baumeister and Tierney (2011) suggest that the students had used up their willpower and perseverance through not eating cookies and had exhausted much of it by the time they had to do the puzzles. This experiment suggests that we have one source of willpower for different challenges – avoiding chocolate and analytical thinking in this case – and that this resource is limited.

This is one hypothesis – that we become mentally weary because we have depleted our internal 'tank' of energy. Another hypothesis is that we feel tired and depleted because our motivation runs out. Brian Resnick's 2018 *Vox* article 'Why your desk job is so damn exhausting' provides more detail on the two theories. We will also look at managing mental energy in Chapter 11. Portable technology (eg tracking devices) means that researchers now have more methods to monitor our levels of fatigue and our emotions at work, and the impact they have on both our productivity and mental well-being.

Resisting chocolate or applying ourselves to hard tasks might seem like obvious demands on our self-control and mental energy, but there's another one that might be less obvious but very relevant to organizations going through change: decision-making. Organizations and leaders need to think about this: when are decision-making meetings scheduled – at the beginning of the day when we still have energy, or later when energy is running low? How many complicated and significant decisions are people being expected to make in one meeting?

As our self-control, energy and motivation get depleted, so we find it harder during the course of the working day to keep control of our emotions. As the day goes on and we have used up our self-control or motivation, we are more likely to get emotional and angry. It's the reason why, if we have had a tough day at work, we can no longer hold back on the chocolate cake that we have done such a good job of resisting all day, or once we walk in the front door at home we are more likely to be grumpy and rude to the people we live with. We have controlled our emotions all day and now our mental energy is depleted.

Once we are in this depleted state, the brain starts to rely on a short cut in terms of making decisions. Teenagers and burnt-out adult brains start to use the 'low road' in terms of decision-making where they rely less on considered 'reflective' thinking and the PFC, and more on what is called 'reflexive' thinking where the amygdala plays a greater role. We will come back to this in Chapter 7 on decision-making.

Another point to note is that people who are in a constant state of anxiety, as many might be if the change programme and uncertainty is prolonged, find it harder to activate the RVLPFC. So, if the mood and climate is one of anxiety in the organization, then people are less likely to have emotional control. This state of anxiety will be contagious and, once again, employees will not be best placed to focus on their work, customers, collaborating or being creative.

Research by Snyder *et al* (2010) illustrates how anxiety impairs our performance even in what would seem to be the simple task of selecting an appropriate verb to go with a noun (such as 'purr', 'meow' or 'lick' with 'cat'). Their research also indicates a link between anxiety and reduced activity in the VLPFC.

## Glucose

Danziger *et al* (2011) have done some fascinating and rather disturbing research: they reviewed over 1,000 decisions made by judges presiding over a parole board in Israel. Their research reveals that anyone going in front of a judge needs to check when the judge last had a break and had a meal – you really don't want them to be deciding your future if it is a while since they last ate. Each judge on average gave parole to one in three prisoners, but these decisions were very different depending on whether they were made before or after the judge had eaten. The judges had two meal breaks during the course of each daily sitting, which meant there were three decision-making sessions each day. Danziger and team noticed that at the start of the sitting, prisoners had about a 65 per cent chance of being given parole. This dropped down to almost 0 per cent during the session, but the chance rose dramatically back to 65 per cent once the judge had had a break and something to eat. Judging is demanding on the brain and the more decisions the judges made, the more glucose was being used up. The less glucose the judges had, the more risk-averse and the harsher they were in their decisions. What is concerning here for all of us is the inconsistency of decision-making which is being influenced, by irrelevant factors. We all need to pause and reflect: how fair are we being to people we are interviewing or whose performance we are reviewing if it is a while since we last took a break and had something to eat?

In their book *Willpower* (2011), Baumeister and Tierney cite two more experiments that reveal the brain's need for glucose. In one experiment, a group of children were told to come into school without having eaten breakfast. One half was then given a healthy breakfast; the other half ate nothing. The children who had eaten learned more and were better behaved (as judged by people who did not know which ones had had breakfast and which ones had not). Then all the children were given a healthy snack and the differences in ability to learn and to behave disappeared. In another experiment glucose levels in adults were measured before and after tasks, such as watching a video where words were flashed up. One group was told to relax and watch the video normally and the other group was told to ignore the words. Levels of glucose remained the same in the relaxed viewers but noticeably dropped in those who had had to ignore the words: a small, simple exercise of self-control can lead to a big drop in the level of glucose. Again, we need to keep this in mind as we plan our work activities and those of the team. Work that needs self-control and discipline should be scheduled earlier rather than later in the day, and food breaks need to be considered.

## People with emotional control tend to fare better in life

Many people are familiar with the experiment conducted by Walter Mischel of Columbia University in the 1960s (Mischel *et al*, 1989). Mischel at the time had young daughters and was fascinated by the difference in their ability to control their behaviour as they grew older. This led to the marshmallow experiment. Mischel has done this numerous times but the key parts are that children who are around four years old are brought into a room: there is a marshmallow on the table and the child is told they can have the marshmallow now or, if they wait until the researcher returns, they can have two. The child is left alone in the room. Mischel has filmed some of the children, watching the different tactics they use to aid their self-control. Some succeeded; others couldn't resist. Where Mischel's research became even more interesting was in following them through life: those children who were able to apply more self-control in the experiment were more likely to get better grades in school, earn higher salaries, were less likely to become overweight and to have problems with drug abuse. Again, there have been critiques of this research, some questioning whether the children who showed less self-control came from backgrounds where the children had learned that adults can be unreliable, in which case it made good sense to take the one marshmallow immediately; in their experience the proverbial bird in the hand being better than two in the bush.

## A model based on social cognitive neuroscience to help manage our emotions

James Gross, Professor at Stanford University with a specific interest in emotion regulation, has developed a model of the stages and options we have in managing our emotions (1998). He has developed this further with Kevin Ochsner, Professor of Social Cognitive Neuroscience at Columbia University (Ochsner and Gross, 2007).

### Five stages to calm our brains and deal with emotions

The five stages identified by Gross and Ochsner can be separated into two groups: those tactics we can try before the situation arises and those that we can apply during the event. They can also be divided into actions where we change the external environment and actions that involve changing ourselves internally and how we perceive the situation. The model should be applied to something you know will provoke emotions that you'd rather not experience: it might be getting very nervous during public speaking or having to work with a colleague whom you don't particularly like. It might be useful to have a recent unpleasant event in mind and see which of these five approaches would have helped you to stay calm and in control.

*Changes you can make before the event:*

1 Situation selection (external) – this is about choosing situations that will spark emotions that we would like to have and avoiding situations that cause emotions we don't want. For example, if public speaking makes you really nervous, can you avoid doing it? Is there a better way for you to communicate with people?

2 Situation modification (external) – is there a way in which you can change the situation so that it won't arouse those unwanted feelings or enhance it so that positive feelings are even stronger? For example, if the problem is public speaking, could you ask someone to speak with you? Situation modification might involve enhancing the situation so that it generates even more positive emotions, eg holding a team meeting but making sure that it's in a room with plenty of natural light, that there are refreshments, and that the temperature is just right, in order to bring out the best in the team.

*Changes you can make during the event:*

3 Attention deployment (internal) – this is not about changing the situation or environment but about changing how we deal with it internally: how we perceive the situation. Distraction is one technique: using public speaking again as our example, you agree to make the presentation, but you decide to focus on just one friendly-looking person in the audience so that you feel less nervous. Mischel said a lot of children used attention deployment as a tactic in the marshmallow test – they looked elsewhere so that they did not have to see the tempting marshmallow. We sometimes use this tactic with pain when we choose to focus on something else other than the pain. Parents often use it with children: telling them a story, for example, when the child is feeling upset or hungry. Research by Erman Misirlisoy *et al* (2015) shows that attention redeployment (focusing on something in the external world or focusing on a deliberate, voluntary movement) can help people with Tourette's syndrome reduce the frequency of their tics. It can be very useful in the organization too, asking people to focus on something that elicits positive emotions: thinking about the benefits of the change programme, for example. In many organizations going through difficult times, employees have decided to get involved in fund-raising for a charity: focusing on doing something constructive for others.

4 Cognitive change (internal) – the meaning that we give to a situation affects how we feel about it; seeing the situation differently leads to different emotions. In using cognitive change we are trying to change

the emotional significance of the event. This is commonly known as re-appraisal. There are several means of doing this:

- We can accept the emotion but recognize that it is just an emotion that will pass (you are not your emotions), ie we try to separate ourselves from the emotion: see the section on mindfulness in this chapter for more on developing this ability.

- Reframing: reinterpret events so that we look at the event more positively – choose to interpret it in a way kinder to yourself and to others. An example I had of this was a client who had been on holiday for a week, returned to work, and found that a colleague of whom she was not particularly fond had called a supplier and told them to stop working on my client's project. My client was livid and was about to tell her colleague exactly what she thought of her. We talked and I asked her to pause – did she know why the colleague had contacted the supplier? Perhaps there was a sound reason that my client was unaware of? Perhaps she had been told by someone more senior to stop the supplier? There might have been a perfectly good and reasonable explanation and perhaps the colleague had acted with the best of intentions.

Here's another example, shared in a workshop by someone who suffers from stress: he had a somewhat unpredictable commute to work and when there were delays on the train, he got particularly anxious. He had learned to reappraise this by seeing a delay on the train as a bit more 'me time' to read a book, for example. This reframing had worked and he felt less stressed. Similarly, if we feel nervous before public speaking, it can help to acknowledge that feeling a bit nervous is a good thing as it is energizing and sharpens up our thinking. Having emotions and bodily sensations is inevitable. Reframing can make them useful rather than problematic.

This technique is about realizing that there is often more than one way to view an event: we can reinterpret the situation to take a different perspective and so reduce negative emotions.

An important point to note about reappraisal is that it reduces activity in the amygdala and in the insula: they no longer 'see' an emotionally-arousing event. You have enabled your brain to see the situation differently and you have removed the threat. This means that our abilities to stay focused, think clearly and remember are greatly enhanced – important abilities for employees, especially when going through change.

- Distance yourself: viewing yourself in the third person and even talking about yourself in the third person reduces emotions because it puts some distance between ourselves and the events. Coaches sometimes use this technique in sessions with their clients – asking the client to move to another chair and look back at themselves, so to speak, so that they can try to see things more objectively and less emotionally. Looking at yourself as a third person is an especially effective method for dealing with difficult past events.

  Timothy Wilson, in his book *Redirect* (2011), describes research where people who had had a difficult past experience, perhaps with an aggressive boss, were asked to think about that experience again. The most successful approach (steady blood pressure, fewer negative emotions) was the 'step back and ask why' approach, ie those people who distanced themselves from the event and thought about *why* the event had happened, rather than focusing on the emotions, fared much better. As Wilson says, it is best not to recount the event but to step back, reconstrue it and explain it. He gives the example: "'I see now that my boss's anger had more to do with his impending divorce than with anything about me, and now that I think about it, I have to admit that I could have done a better job on that report.'"

5 Response modulation – this one occurs later in our efforts to manage our emotions, after or during the emotional experience – and it includes dealing with our physiological or behavioural responses to the situation. Exercise and relaxation are examples. Suppression is a common form, and we frequently use this at work where we want to look calm, even when we might be feeling quite unhappy or worried about something. For example, we may hide our anger at what the boss has just asked us to do, or our anxiety before we have to speak in front of a group of people, or our worries about the personal implications of a major restructuring. But suppression comes at a cost – cognitively and socially. Research by Richards and Gross (2006) shows that suppression leads to worse memory for what happened during the period of suppressing, as much as distracting yourself: so if you are doing this during a meeting, your recall of what went on in the meeting will be poor. Suppression costs us in many more ways too (Butler *et al*, 2003):

   - Suppressors are distracted because they are trying to do two things at once – suppress and perhaps hold a conversation (and despite what we like to think, we cannot multi-task).

- The blood pressure of the person suppressing the emotion goes up.
- The blood pressure of the 'partner' goes up too – they pick up on the emotion of the suppressor.
- When people are not well acquainted, suppression also gets in the way of developing a good relationship and in building rapport.

Another point to note: whereas reappraisal reduces activity in the amygdala and the insula, suppression increases it. Sometimes we might need to use suppression in the short term to get us through an event, but it is not a beneficial technique in the long term. When I talk about Butler *et al*'s work on suppression, the point that particularly concerns people is that they are not just increasing their own blood pressure, but that of their co-workers too.

## Managing our emotions – mindfulness

Meditation and mindfulness have been around for thousands of years, but there has been a noticeable increase in interest in them in the West in recent years. Put simply, mindfulness is a mental discipline that enables us to become more aware of our own thoughts, to be able to observe them and choose which thoughts to follow. Practised regularly, it enables us to observe our emotions non-judgementally and enables us not to be overwhelmed by them, hence its inclusion in this chapter. Mindfulness is about tuning in and improving attention – it is not a form of distraction. It enables us to switch off that constant narrative that we have in our heads. It is a well-researched area in neuroscience (over 3,000 papers) and, although some of the studies are on a small base and need the findings to be replicated, some of the findings are pretty compelling:

- improved attention and control of stress (Tang *et al*, 2007);
- physical changes in the brain – increased cortical thickness in regions involved in:
  - attention;
  - self-awareness;
  - emotion regulation;
  - learning and memory (Lazar *et al*, 2005; Hölzel *et al*, 2011);
- reduced amygdala activity (Way *et al*, 2010);
- reduced fatigue and anxiety (Zeidan *et al*, 2010);

- improved working memory and executive functioning (Zeidan *et al*, 2010);

- lower neuroticism and increased empathy (Krasner *et al*, 2009);

- higher emotional intelligence (Baer *et al*, 2004);

- slowed ageing in the brain and improved attention (Pagnoni *et al*, 2007);

- improved decision-making through recognition and reduction of cognitive biases (Hafenbrack *et al*, 2014).

The last bullet point refers to research conducted by Andrew Hafenbrack (2014) at INSEAD with his colleagues Zoe Kinias, Assistant Professor of Organizational Behaviour at INSEAD, and Sigal Barsade of Wharton School at the University of Pennsylvania. They were looking at the problems we have in cutting our losses: we carry on with failing projects for too long because of what is known as the sunk-cost bias. This bias means that we find it difficult to let go of a project because we have already invested so much time and effort in it and we don't want to write off that time; but staying with it is just throwing good money after bad. This bias is reflected in the tendency people have to hold on to stocks and shares when they have fallen from their original purchase price, when it would be wiser to sell them. Hafenbrack *et al*'s research showed that people who practised mindfulness for 15 minutes a day were more likely to make rational decisions. The participants were more capable of looking objectively at the current situation and information and were able to let go of the concerns and emotions that exacerbate the 'sunk-cost bias' and that lead us to hang on to bad decisions.

Most of the research into mindfulness to date has looked at the impact on individuals. There's a great deal of research being conducted amongst school children to see if it has an impact on behaviour and learning. Mindfulness is also being taught to prisoners to see if it helps to reduce anger and reoffending. Many organizations are beginning to offer mindfulness classes to employees as they recognize the benefits both to employees and to the organization. Transport for London (TfL) is one example of this. They introduced mindfulness as part of a stress management course. TfL said that, as a consequence, absenteeism caused by stress, anxiety and depression dropped by 71 per cent.

One of the pioneers of mindfulness at Google is Chade-Meng Tan, who has the great job title of Jolly Good Fellow. In his book, *Search Inside Yourself* (2013), he describes the benefits of mindfulness from a leader's point of view, stating that it gives leaders choice over how they respond to events, and that this choice gives them self-control and power.

## Control helps reduce cortisol

When people are feeling stressed, a perception of having some control over what is happening reduces cortisol – we will look at this further in Chapter 8.

## Labelling emotions reduces the intensity

A brain scanning study shows that naming emotions, or 'labelling' them, reduces the intensity of the emotion. Previous studies had found that naming an emotion reduces its impact but this study, led by Matt Lieberman of UCLA, showed what happens in the brain as we do so (Lieberman *et al*, 2007). The act of naming the emotion increases activity in the right ventrolateral prefrontal cortex (RVLPFC) and reduces activity in the amygdala. The RVLPFC increases its activity to dampen down the emotions triggered by the amygdala. However, it's not clear whether this happens for both positive and negative emotions, as 80 per cent of the faces in the study had expressions of anger or fear, while only 20 per cent showed happiness or surprise, so this study only really tells us about negative feelings.

Research conducted by Fan *et al* in 2019 revealed that expressing an emotion on Twitter, be it negative or positive, can attenuate it. Fan and team tracked the evolution of tweets of 74,487 Twitter users and analysed the content of tweets before and after the tweet where they explicitly commented on their emotion. They found that negative emotions build up more slowly than positive ones, and then sharply fall to previous levels. In a review article, Torre and Lieberman (2018) suggest several reasons why affect labelling might have an impact: it could be a form of distraction, it could make the emotion feel less nebulous and therefore reduce uncertainty and/or labelling could provide a 'symbolic conversion' where putting the emotion into words creates a distancing effect and so reduces the emotion.

# What can we do? Solutions and examples of what other leaders have done

As we have explored in this chapter, emotions have a significant impact on how we perceive organizational change and on our ability to stay focused. People leading change have a responsibility to manage their own emotions and to help employees manage theirs. The last section provided insights from a neuroscience perspective on why and how we can control our emotions;

this section looks at what practical interventions we can make. It's useful to remember Carol Dweck's growth mindset (see Chapter 4) at this point too: if we believe that we can learn and improve our ability to manage our emotions in a constructive way, then we are more likely to succeed.

## Managing our own emotions

As we have seen throughout this chapter, the emotions of leaders have a significant impact on others. They need to find the right balance between being empathetic with employees while not getting unduly influenced by others' negative emotions. When leaders get too caught up in others' negative emotions, their amygdala will be overactivated.

### Practising mindfulness

Until recently mindfulness and meditation would be seen by many as something practised by Buddhists or hippies. Not everyone is comfortable with the idea but it is useful to be aware of it. As mentioned in *Managing our emotions – mindfulness* in the Science section, organizations such as Google, Dow Chemical, the UK's Houses of Parliament, GlaxoSmithKline and Transport for London have introduced it. The research is building and just about every neuroscientist I have met says they practise mindfulness. In fact, one neuroscientist says that as we get older there are two key things we need to do to keep our brains functioning well: take exercise and practise mindfulness (others add a third: learn to play a musical instrument or a foreign language – keep pushing your brain to learn). When we ask participants at the end of applied neuroscience workshops what actions they will take, many people say they would like to practise mindfulness. There are lots of books, courses and apps available to help people get started and there are different approaches to mindfulness. One of the most well-known techniques is to focus on the breath. Here's a brief guide:

**Focus on the breath**
Sit in a chair, slightly more upright than you usually would (mindfulness is about being relaxed, but focused – it's not about drifting off), put both feet on the floor and rest your hands in your lap. If you don't feel comfortable on a chair, lie on the ground. Close your eyes and notice how you are feeling throughout your body. When you are ready, focus on your breath – inbreath, outbreath – just keep doing that. Thoughts will come, but gently let them go and focus back on your breath: inbreath and outbreath. Perhaps start with one minute and try to build up each day. Neuroscientists are researching

how long you need to practise for at a time to make a significant difference: at least 12 minutes is one suggestion. The key is to keep doing it each day – it gets better with practice and you are likely to experience more of the benefits listed in the section *Managing our emotions – mindfulness*. Some research mentioned in the Science section suggests people find a benefit after just five days of practising. In addition to the 12+ minutes of mindfulness, build it into your day. Some people put a coloured dot somewhere on their desk and every time they see it, they focus on their breath for, say, three breaths. If you drive to work, next time you are stopped by a red light, rather than seeing it as a frustration, see it as an opportunity to focus on your breath: it's an easy way to turn a negative into a positive. Another manager said every time she opens the fridge door at home or at work, she pauses. A suggestion from another manager was to start each meeting with a pause so that people can be fully 'present' in the meeting, rather than ruminating on the last meeting they have just left. Some coaches start their coaching sessions with a 'minute's mindfulness' so their client is ready for the coaching session and less distracted by the demands of work. A senior manager at an accountancy firm says she finds a minute's mindfulness very useful in calming her down and helping her to reprioritize what she needs to get done in her busy day. A leader in a transport company, who recognizes that he gets angry and frustrated very quickly, said practising a few breaths of mindfulness helps him to get rid of the anger and focus back on the meeting.

### Five things

Another simple mindfulness exercise is 'five things': sit still and look at five things, listen out for five different sounds, feel five things, smell five things and taste five things – for the latter two you might need to think back to earlier in the day. This is a simple technique that calms the mind, brings you back into the present and helps the mind to connect with the body.

## Dopamine and oxytocin

In Chapter 4 we looked at some of the things we can do to generate dopamine. This can be a useful thing to do when we are feeling negative: reminding ourselves of past successes, breaking large tasks into smaller ones so that we feel we have achieved something each day.

Oxytocin is a powerful hormone that can also stimulate positive feelings and plays a role in helping people to bond – it is particularly strong in new mothers and their babies. A hug generates oxytocin which explains why group hugs can feel good. I have seen a facilitator use it in a workshop but it made some people uncomfortable, so use with caution! Paul Zak, a 'neuroeconomist', writes extensively on the role of oxytocin in trust (2013).

# Helping others manage their emotions

One of the most important roles for those leading change is to help others manage their emotions. Managers in particular need support as research (Kets de Vries and Balazs, 1997) suggests that they often emerge from change programmes, particularly downsizing, feeling more depressed than others. This might well be due to the difficult role they have to play of communicating the corporate messages, managing the team, while having little or no control over the decisions that have been made – a very difficult place to be. Here are some examples of what other organizations have done.

## Support for managers

Before offering help to leaders and managers it is worth reflecting and analysing where interventions will make the most impact. Budgets, resources and time are limited, so any support for managers needs to be focused on the right people. It is useful to identify who the 'catastrophizers' are – people who take a negative view of the world – as they might have a particularly negative impact on other employees. Organizations also need to identify who the real influencers are – who are the people that others listen to or who have many points of contact in the organization? One organization literally mapped out who were the 'go to' people in the organization (and it did not follow the hierarchy). It's important to know how these people view the changes and what messages and emotions they are sharing.

### Practical and cathartic workshops

A merger of two government departments was followed by office closures and downsizing. Managers were in that difficult position of having to try to communicate organizational messages and keep teams working, while at the same time having little control over the loss of jobs. The government department offered managers two-hour workshops. The purpose was in part to share knowledge and ideas, but it was also intended as a cathartic session – a safe place for managers to vent their feelings. Numbers in each group were limited to around 15 people. The agenda was kept short so that there was plenty of time for discussion:

1 What issues are you facing?
2 The impact of change and how to support your team.
3 Engaging people in difficult times – helping people to let go of the past, managing people in 'limbo', setting up new teams.
4 Ongoing support for managers.

An important aspect of these workshops was giving managers a safe place to vent and 'label' their emotions. As we know, emotions are contagious, and it was better to let managers deal with their emotions with their peers and experienced facilitators than to leave them suppressing their emotions (which, as we know, does not work) or letting them vent in front of their teams. The workshops received very positive feedback, not least because the fact that facilitators were travelling to the managers was a sign of respect from the organization to managers and a recognition of the difficult and influential role managers play during change.

### Focus on performance during change

Shortly after the global banking crisis, one of the UK banks that was down-sizing was concerned that employees should be equipped to perform at their best during these turbulent times. Leaders and managers from across Europe were offered half-day workshops. These were in small groups to enable leaders to talk openly and to apply the learning to their team. The purpose was to build resilience and maintain performance. The agenda:

1 What challenges are you facing?
   - Setting a context for the workshop.

2 What have you observed from your experience of change to date?
   - People drew a 'timeline' of their lives, indicating the highs and lows, and talked about what got them through difficult times in the past. Everyone said they had faced tougher situations than the current one – this helped to build a sense of resilience.

3 The climate that change creates – what should we expect to see going on?
   - This got people talking about how people were responding to change and showed that many of the reactions were normal and not peculiar to this bank.

4 Resilience.
   - A questionnaire to assess their resilience and to get them to reflect on the importance of sleep, exercise, social connections and healthy food.

5 What neuroscience can teach us about performance.
   - This included some of the content covered in Chapters 3 and 4 and gave people practical things they could do to keep people focused and positive.

6 Dealing with Transition (based on William Bridges' model – see more on this in Chapter 9).

**7** What will you do?

- Participants discussed what they would apply and when. These actions were followed up by the HR team.

## Enabling employees to talk

As we saw in the last chapter on the social brain and in this one, being able to talk about emotions (but not wallow in them) and feeling listened to can help people deal with negative emotions. Sometimes managers find this uncomfortable because there are not necessarily any answers, or not ones that people want to hear. Some training for managers before such sessions can be useful, eg teaching them about active listening, summarizing and paraphrasing. Listening empathetically does not mean that the manager agrees with what is being said but it can be helpful to managers to know where to let people talk and where clarification and correction of misinformation is needed.

## Reappraisal

Having the ability to help others reframe and reappraise is a useful skill. As we saw in the Science section, reappraising – having the ability to take different perspectives on what is going on – is a 'brain-friendly' thing to do, dampening down the amygdala. Reframing during change is particularly important as the threat response means that we see the work environment through a more negative lens. Reframing means employees are more able to see not just the threats in change but the opportunities too. Perspective-taking improves with practice. Some examples of this might be getting employees to look at the outcomes of the changes from a customer's perspective – what will be better for them? Restructuring often means that some employees are asked to take new roles and they become anxious about having to learn new skills and about not being competent enough to do the new job. A useful course of action from the manager would be to ask the employee about when they have successfully learned new skills in the past, and to focus on why they think they have been selected for this new role, and what experience and skills they already possess that might be useful in the new role.

### Appreciative Inquiry

One very useful technique to use with teams is Appreciative Inquiry (AI). So often in organizations we tend to focus on what is wrong and what needs fixing, but AI forces us to take a different perspective by asking what's working well and what can we learn from it? The very fact of asking people to focus on the positive is an affirmative experience and pushes people towards a 'reward response', which is a more positive, creative place for the brain to be. The concept was developed by David Cooperrider and Suresh Srivastva.

AI uses a four-stage process (Cooperrider, 2008):

1 **Discover**: the identification of organizational processes that work well.

2 **Dream**: the envisioning of processes that would work well in the future.

3 **Design**: planning and prioritizing processes that would work well.

4 **Deliver**: the implementation of the proposed design.

A good way to start is to ask the group to initially work in pairs and, depending on the nature of what you are focusing on, ask them to interview each other about a real example of when they have done something well. Equip the interviewer with questions such as:

- Tell me about the situation and why it worked so well.
- What did you do that made it work so well?
- What are you proud of?
- What positive impact did it have on others?
- What did you learn?
- How might we apply this learning to the organization?

## Distancing

The 'step back and ask why' approach described in Timothy Wilson's book (2011) is a useful way to help employees 'move on' and dwell less on negative past events. Distancing, as with many of the techniques that help us control our emotions, gets better with practice. All of us, but particularly leaders, need to find strategies that work for us. They need to build these techniques up ahead of time so that, when work is particularly stressful or wearing, they are already well practised in the techniques.

# Summary of key points from this chapter

- Becoming more aware of our emotions means that we are better able to choose how we respond to situations and can learn to avoid what is less helpful about them.

- Emotions are a crucial part of decision-making: people with damage to parts of their brains that process emotions can find it very hard to make even the simplest of decisions.

- Managing our emotions is important because:
  - Emotions such as fear, anger and anxiety affect our ability to perform well.
  - We are subject to a great deal of stressors in the workplace.
  - Stress affects us in the short term, and in the longer term is physically damaging and destroys brain cells.
  - Emotions are contagious, especially those of leaders.

- We can get better at managing our emotions, and neuroscience reveals insights as to how we can do this.

- Critical systems in processing emotions include the amygdala (fear), hippocampus (memory), hypothalamus (coordinator and triggers response to stress), the insula (internal stimuli) and the RVLPFC (brain's braking system).

- Constant stress and anxiety lead to an overactive and oversensitive amygdala that is more easily activated.

- Causes of workplace stress include high demands on people, especially when they have little control, boredom, continuous layers of change and uncertainty.

- There are differing theories as to why we become mentally tired: one theory is that we have one 'tank' of self-control that we draw on, whether refraining from eating marshmallows or applying ourselves to working on difficult analytical problems. Another theory is that we lose motivation.

- Glucose levels influence our decision-making and ability to think and learn.

- People with emotional control tend to fare better in life.

- When managing our emotions we have choices as to how we do that (Gross, 1998; Ochsner and Gross, 2007):

  1 Situation selection – avoid situations that trigger negative emotions.

  2 Situation modification – change the situation so that it is less emotional.

  3 Attention deployment – focus on something different to dampen negative emotions.

  4 Cognitive change – look at the situation differently, take a different perspective – reframe it; put some psychological distance between yourself and the event.

5 Response modulation – deal with your responses. Exercise can be a positive way to deal with negative emotions after the event. But beware suppressing your emotions – it has negative consequences for you and for others and activates the amygdala.

- Mindfulness is a well-researched technique for managing our emotions.
- The act of labelling negative emotions reduces them.
- A perception of control reduces cortisol.

## What can we do?

- Manage our own emotions.
  - Get the balance right between empathy and not getting too caught up in others' emotions.
  - Be aware that we can get better at managing our emotions (growth mindset).
  - Use techniques such as mindfulness so that we learn to pause and respond, not just react in the heat of the moment.
  - Use activities that activate the reward network, eg thinking about past achievements.
  - Talk about emotions (but don't wallow) – naming negative emotions reduces them.
- Help others manage their emotions.
- Workshops for managers that allow them to vent and then act.

  - Resilience workshops for managers that equip them to help others focus and perform.
  - Listen, empathize, allow people to label emotions but not to ruminate.
  - Equip employees to reappraise and take different perspectives to reduce amygdala activity.
  - Use Appreciative Inquiry as a technique to get people to focus on the positive and the practical and to apply this learning to the changes.
  - Help employees to distance themselves from a painful event and analyse why it happened.

# Reflections and planning

We know that emotions are contagious, especially those of leaders:

- Are leaders in your organization aware of the impact of their emotions?
- Are leaders aware of the many different ways in which employees pick up conscious or unconscious emotions from them?
- What processes, support and training do you have in place to help leaders manage their emotions so that they can create the kind of climate that helps employees?
- Who are the real influencers? How do they feel about the changes? What emotions are they conveying to others?
- Change programmes often focus on facts and logic. Do you discuss the emotional aspects of change with leaders? Are they comfortable discussing the emotional aspects with employees?

We know that self-control is limited:

- Where are you taking this into account in the change management process?
- When are decision-making meetings planned?
- How many decisions are on the agenda?
- How do you ensure glucose levels are at the right level for sound decision-making?
- Are there ways in which you can reduce the demands on employees' self-control?
- What's the working environment like? In what ways might it drain people's self-control? Do you know?
- What's the culture like? Do people feel they need to hide their emotions, suppress them and look resilient? If so, are people really able to think and focus at their best?
- Might attention deficit trait be an issue in your organization – are executives turning into frenzied underachievers?

We know that negative emotions impede people's ability to focus, think and collaborate:

- How do people see the changes? As a threat? As an opportunity? Both?
- Are there catastrophizers in the organization? If so, how are you managing them?
- What is the organization doing to help people reframe the changes?
- How able are managers at helping their teams to reinterpret, reframe, and distance themselves from painful emotions? Do they have the skills?
- How well balanced are your communications about the changes? Do they help people to focus on the positive outcomes? Are people aware of the successes?
- How are you ensuring that people feel valued and respected?
- What are their causes of concern? Do you know?

We know that managers are in a particularly difficult place in change programmes – responsible for their teams but often with little influence over the decisions:

- What support do you have in place for managers?
- How well do you communicate with them about the changes? What means do you have to listen to them and the challenges they are facing in supporting the change?
- Do they have a safe space to vent and to ask their questions?

We know that cortisol is damaging in the long term for the body and brain:

- What stress reduction programmes are in place?
- How many sickness days are due to workplace stress?
- Are there ways in which you can reduce the stressors for employees?
- Do leaders know that a little stress and challenge is good, but too much and the PFC closes down?
- Boredom is a cause of stress. Are some jobs boring and repetitive in your organization? If so, what can you do to relieve this boredom?
- A perception of control reduces cortisol:
  - How do people feel about the changes: do they feel like victims of the changes?
  - Do employees have any influence over the changes?
  - What are you doing to give people some control over the changes?

# References and further reading

Baer, R A, Smith, G T and Allen, K B (2004) Assessment of mindfulness by self-report: the Kentucky inventory of mindfulness skills, *Assessment*, **11** (3), pp 191–206

Baumeister, R F and Tierney, J (2011) *Willpower: Why Self-Control is the Secret to Success*, Penguin, London

Bridges, W (1991) *Managing Transitions*, Perseus Books, New York

Butler, E A *et al* (2003) The social consequences of expressive emotion, *Emotion*, **3** (1), pp 48–67

CIPD (2018) [accessed 4 January 2019] Health and well-being at work [Online] https://www.cipd.co.uk/knowledge/culture/well-being/health-well-being-work

Cohen, S, Janicki-Deverts, D and Miller, G E (2007) Psychological stress and disease, *JAMA*, **298** (14), pp 1685–87

Cooperrider, D L (2008) *Appreciative Inquiry Handbook*, Berrett-Koehler, San Francisco CA

Damasio, A (2006) *Descartes' Error: Emotion, Reason, and the Human Brain*, Vintage Books, New York

Danziger, S, Levav, J and Avnaim-Pesso, L (2011) Extraneous factors in judicial decisions, *Proceedings of the National Academy of Sciences of the United States of America (PNAS)*, **108** (17), pp 6889–92

Ekman, P, Sorenson, E R and Friesen, W V (1969) Pan-cultural elements in facial displays of emotions, *Science*, **164** (3875), pp 86–88

Fan, R, Varamesh, A, Varol, O, Barron, A, van de Leemput, I, Scheffer, M and Bollen, J (2019) The minute-scale dynamics of online emotions reveal the effects of affect labelling, *Nature Human Behaviour*, **3**, pp 92–100

Friese, M, Loschelder, D D, Gieseler, K, Frankenbach, J and Inzlicht, M (2018) Is ego depletion real? An analysis of arguments, *Personality and Social Psychology Review*, https://doi.org/10.1177/1088868318762183

Gallese, V and Goldman, A (1998) Mirror neurons and the simulation theory of mind-reading, *Trends in Cognitive Science*, **2** (12), pp 493–501

Ganzel, B L *et al* (2008) Resilience after 9/11: multimodal neuroimaging evidence for stress-related change in the healthy adult brain, *NeuroImage*, **40** (2), pp 788–95

Gross, J J (1998) The emerging field of emotion regulation: an integrative review, *Review of General Psychology*, **2** (3), pp 271–99

Hafenbrack, A C, Kinias, Z and Barsade, S G (2014) Debiasing the mind through meditation: mindfulness and the sunk-cost bias, *Psychological Science*, **25**, pp 369–76

Hallowell, E M (2005) Overloaded circuits: why smart people underperform, *Harvard Business Review*, **83** (1), pp 54–62

Herbert, T B and Cohen, S (1993) Stress and immunity in humans: a meta-analytic review, *Psychosomatic Medicine*, **55** (4), pp 364–79

Hölzel, B K *et al* (2011) Mindfulness practice leads to increases in regional brain gray matter density, *Psychiatry Research: Neuroimaging*, **191** (1), pp 36–43

Jones, A C and Josephs, R A (2006) Interspecies hormonal interactions between man and the domestic dog, *Hormones and Behaviour*, **50**, pp 393–400

Kets de Vries, M F R and Balazs, K (1997) The downside of downsizing, *Human Relations*, **50** (1), pp 11–50

Krasner, M S *et al* (2009) Association of an educational program in mindful communication with burnout, empathy, and attitudes among primary care physicians, *JAMA*, **302** (12), pp 1284–93

Lazar, S W *et al* (2005) Meditation experience is associated with increased cortical thickness, *Neuroreport*, **16** (17), pp 1893–97

Lieberman, M D *et al* (2007) Putting feelings into words: affect labeling disrupts amygdala activity in response to affective stimuli, *Psychological Science*, **18** (5), pp 421–28

McEwen, B S, Weiss, J M and Schwartz, L S (1968) Selective retention of corticosterone by limbic structures in rat brain, *Nature*, **220** (5170), pp 911–12

Mehta, P H, Jones, A C and Josephs, R A (2008) The social endocrinology of dominance: basal testosterone predicts cortisol changes and behavior following victory and defeat, *Journal of Personality and Social Psychology*, **94** (6), pp 1078–93

Mischel, W, Shoda, Y and Rodriquez, M I (1989) Delay of gratification in children, *Science*, **244** (4907), pp 933–38

Misirlisoy, E, Brandt, V, Ganos, C, Tübing, J, Münchau, A and Haggard, P (2015) The relation between attention and tic generation in Tourette syndrome, *Neuropsychology*, **29** (4), pp 658–65

Monk, C S, Telzer, E H and Mogg, K (2008) Amygdala and ventrolateral prefrontal cortex activation to masked angry faces in children and adolescents with generalized anxiety disorder, *Arch Gen Psychiatry*, **65** (5), pp 568–76

Morris, J S, Öhman, A and Dolan, R J (1998) Conscious and unconscious emotional learning in the human amygdala, *Nature*, **393** (6684), pp 467–70

Ochsner, K N and Gross, J J (2007) The neural architecture of emotion regulation, in *Handbook of emotion regulation*, ed J J Gross, pp 87–109, Guilford Press, New York

Pagnoni, G and Cekic, M (2007) Age effects on gray matter volume and attentional performance in Zen meditation, *Neurobiology of Aging*, **28** (10), pp 1623–27

Resnick, B (2018) [accessed 18 November 2018] Why your desk job is so damn exhausting [Online] https://www.vox.com/science-and-health/2018/9/5/17818170/work-fatigue-exhaution-psychology

Richards, J M and Gross, J J (2006) Personality and emotional memory: how regulating emotion impairs memory for emotional events, *Journal of Research in Personality*, **40** (5), pp 631–51

Rizzolatti, G and Craighero, L (2004) The mirror-neuron system, *Annual Review of Neuroscience*, **27**, pp 169–92

Sapolsky, R M (1998) *Why Zebras Don't Get Ulcers*, WH Freeman & Co

Sherman, G D, Rice, L K, Jin, E S, Jones, A C and Josephs, R A (2016) Sex differences in cortisol's regulation of affiliative behavior, *Hormones and Behavior*, http://dx.doi.org/10.1016/j.yhbeh.2016.12.005

Snyder, H R, Hutchinson, N, Nyhus, E, Curran, T, Banich, M T, O'Reilly, R C and Munakata, Y (2010) Neural inhibition enables selection during language processing, *PNAS*, **107** (38), pp 16483–88

Sy, T, Côté, S and Saavedra, R (2005) The contagious leader: impact of the leader's mood on the mood of group members, group affective tone, and group processes, *Journal of Applied Psychology*, **90** (2), pp 295–305

Tan, C-M (2013) *Search Inside Yourself: The Unexpected Path to Achieving Success, Happiness (and World Peace)*, HarperCollins, New York

Tang, Y *et al* (2007) Short-term meditation training improves attention and self-regulation, *Proceedings of the National Academy of Sciences of the United States of America (PNAS)*, **104** (43), pp 17,152–56

Torre, J B and Lieberman, M D (2018) Putting feelings into words: affect labelling as implicit emotion regulation, *Emotion Review*, **10** (2), pp 116–24

Way, B M *et al* (2010) Dispositional mindfulness and depressive symptomatology: correlations with limbic and self-referential neural activity during rest, *Emotion*, **10** (1), pp 12–24

Wilson, T D (2011) *Redirect: Changing the Stories we Live By*, Penguin Books, London

Zak, P J (2013) *The Moral Molecule: The New Science of What Makes Us Good or Evil*, Corgi Books, London

Zeidan, F *et al* (2010) Mindfulness meditation improves cognition: evidence of brief mental training, *Consciousness and Cognition*, **19** (2), pp 597–605

# Decision-making and bias

07

For obvious reasons there is a great deal of interest in understanding how we make decisions. For organizations, it is crucial to get it right. We can all name companies that have now declined or disappeared through making poor decisions – Nokia, Eastman Kodak, Blockbuster, to name just a few. Closer to home, we all experience good and bad decision-making in the organizations we work with and we wonder how people could have thought those faulty decisions were right. For centuries the belief was that our decision-making was largely rational, but then research by, amongst others, Daniel Kahneman and Amos Tversky in the 1970s showed that in certain circumstances our decision-making is far from rational. In more recent years, behavioural economics, and books such as *Nudge* (2009) by Richard Thaler and Cass Sunstein and *Influence* (2001) by Robert Cialdini have changed our understanding of how people make choices and how they can be influenced. Governments have adopted some of this thinking to help citizens make sensible choices about health, pensions, and energy conservation. Examples of this include requiring citizens to opt out of organ donation rather than actively having to opt in, reminding people that the norm is for people to complete their tax returns on time, arranging for employees to be automatically enrolled in a pension plan, and putting posters up in neighbourhoods stating how many families no longer use their car to take children to school.

Decision-making during change can be particularly difficult: there's a need for swift decisions and yet we are often surrounded by uncertainty. We know from earlier chapters that when we are surrounded by uncertainty we are in a 'threat state', where we see the work environment as more threatening than it really is. The decisions made early in a change programme are particularly important as they set the direction for the changes, and employees are particularly vigilant, wanting to understand where the changes are heading and how they personally might be affected. Making the wrong decision can throw a change programme off course. In one global financial services company, there was too little involvement of influential leaders early in the changes: they felt excluded and alienated by the proposed changes. This lack of ownership on their part meant that they resisted much of what was proposed. In a merger in the United States, the Human Resources team communicated quickly about the severance deal but too slowly about the relocation offer – employees took this as a message that they should leave. The result was that far too many Research and Development people left, leaving the newly-merged company under-resourced in a key area. In another global change programme, one leader refused to accept the proposed change process. He expected the CEO to deal with him quickly, but the CEO hesitated and, one by one, other leaders said they did not want to cooperate either. The result was that the change programme unravelled and it took many more months to implement, as the CEO had to embark on persuading his leadership team to work with him.

It's also important to remember that biases have a constant impact on the decisions we make and the danger is that we are largely unaware of them. We think we view the world objectively but we are just looking at a construct created by our brains. As the concave and convex circles in Figure 2.2 in Chapter 2 showed, our brain interprets on our behalf without us knowing. We construct a picture of what we think is reality, but it's not. Some biases are cultural and learned; others are cognitive and shared across cultures and geographies.

The next section takes a look at some of the science behind decision-making and bias.

# The science

The focus of this book is on helping people in organizations introduce change as effectively and as efficiently as they can, so it's worth reflecting on what we have learned so far that has an impact on decision-making. We know from Chapter 3 that uncertainty and ambiguity can send the brain into a threat state. This causes oxygenated blood to flow to those parts of the brain that get us ready for fight or flight and away from the prefrontal cortex (PFC) that plays an important role in thoughtful, considered decision-making. Being in a threat state means that our field of vision literally narrows and we are less able to perceive what is going on around us. We are distracted and more anxious and we have a greater tendency to see the negatives in the workplace. This influences our decision-making and will reduce our ability to find solutions through insight. We looked at what helps the brain reach those moments of insight in Chapter 4.

If Baumeister's work, referred to in Chapter 6, holds true, we have a limited amount of self-control and we use this finite amount for a wide range of decisions – from refraining from eating marshmallows and chocolate, choosing to continue at a difficult intellectual task, to making significant decisions about other people's futures (Baumeister and Tierney, 2011). These feel like very different decisions to us but each one depletes our ability to stay in control and to make more thought-through decisions. If each decision depletes our resources we need to think about when we are making decisions and how many significant decisions we are aiming to make in one meeting. We need to take into account how we structure the agenda of decision-making meetings: where an item comes on the agenda will have an impact on how much thought and focus we will put into that decision. Remember those judges making decisions about parole and who were less likely to take risks the longer the session went on and as their glucose levels dropped. Decision-making depletes our energy and diminishes our ability to regulate thoughts and emotions.

When thinking about decision-making during change we also need to stay aware of Antonio Damasio's work, which demonstrates the importance of emotions in making decisions. As we explored in Chapter 6, we have emotions for a reason: they help to guide our behaviour. Without them, the simplest of decisions can become very protracted. In business, we tend to think that when it comes to decision-making we need to leave emotions out of it and try to be logical and rational, but neuroscience shows that emotions have a role to play and are valuable in nudging us to a decision. Decisions made purely on the basis of reason are not always sound and not all decisions influenced by emotion are flawed.

One useful means of demonstrating to people that reason and emotion play a part in our decisions is the ultimatum game. In this game one player is given some money, and this player must then propose how to divide the money with a second person. The second person has the choice of accepting the offer, and both then get their share, or rejecting the offer, in which case both players get nothing. Rejection of any offer above zero is not economically rational but people will reject the offer if it feels too unfair. In fact, the majority of people will reject offers of less than a third of the total amount of money (Camerer and Thaler, 1995). Fairness matters to us and plays a role in our decision-making. Receiving an unfair offer activates the anterior insula; a fair offer activates the brain's reward system.

# Making decisions

The ability to make good decisions is crucial for any organization, and this is especially true during change when the environment is volatile and uncertain, but the consequences of decisions can be long-lived. This section takes a brief look at some of the models and neuroscience that help explain how we make decisions.

## The diffusion decision model

One of the most well-established models of decision-making is the diffusion model (Ratcliff and McKoon, 2008). As Figure 7.1 shows, for each decision where there are two options, there is an upper and lower threshold, and once we have crossed one of those thresholds we have made our decision. When we are faced with a decision (say, whether to appoint Candidate A or Candidate B to the job) we gather evidence, and this evidence will push us to choose one candidate or the other. We are faced with a speed/accuracy trade-off: if we are under time pressure to appoint one of the candidates very quickly, then we need to accumulate evidence rapidly and be prepared not to gather in all possible information. The problem here would be that we might miss key evidence or base our decision on inaccurate data. So, looking at Figure 7.1, the boundaries or thresholds are closer together when we are in a hurry and further apart when we want to be more accurate. Perfect accuracy would require very wide boundaries and we would have to wait for ever, gathering all possible evidence. This is clearly not practical. It is a trade-off and a useful one to be aware of. Often, the rule of thumb in change programmes is to get on with decision-making and to be decisive, but a diffusion decision model up on the wall or in our heads is a useful

**Figure 7.1**  Diffusion decision model (Ratcliff)

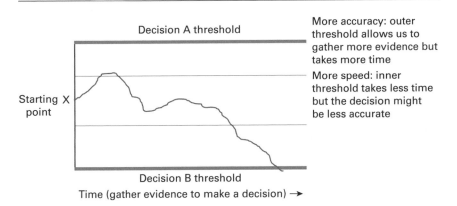

Decision A threshold

More accuracy: outer threshold allows us to gather more evidence but takes more time

More speed: inner threshold takes less time but the decision might be less accurate

Starting X point

Decision B threshold

Time (gather evidence to make a decision) →

reminder of the trade-off we are making and that we are perhaps asking others in the organization to make. Pace is valued in many organizations, and employees feel that if anything the speed of change is increasing. However, if a business rushes employees, those employees will probably make more errors. We know from the Yerkes–Dodson inverted U (see Figure 4.2 in Chapter 4) that some pressure helps us to think more sharply. The challenge is to find just the right balance between speed and accuracy.

## Systems 1 and 2

In his book, *Thinking Fast and Slow* (2012), Daniel Kahneman describes two thinking and decision-making systems which he refers to as Systems 1 and 2. As the book title suggests, System 1 is fast whereas System 2 is slow. The brain is bombarded with information, and most of this information is dealt with by System 1 'below the surface' without us even being aware of it. If you think of learning to drive a car, learning requires System 2 – we need to concentrate and think about what we are doing. Once we have learned to drive, the brain pushes driving into System 1, freeing up our brains to think about other things. System 1 is automatic, reflexive and quick, and, as Kahneman says, it is the system that enables adults to answer 2 + 2 = ? (this is a process that has become automatic with learning). It is also the system that tells us (incorrectly) that the ball costs 10p in Frederick's Cognitive Reflection Test (2005), referred to in Chapter 2. System 2 is slower and takes more effort and is the system that we require when the question is $17 \times 23 = ?$. It takes time to find the answer: we need to work at it. Thinking back to the savannah again: System 1 is the one that will make you jump back or freeze when you see a snake; System 2 will be the one that is slightly slower to tell you that the snake is dead. To conserve energy, the brain largely runs on System 1. As Kahneman says, System 1 is not prone to doubt: it suppresses ambiguity and makes up coherent stories to fit with what we see. System 1 will 'call up' System 2 when more considered thinking is needed. We tend to be much more aware of thinking and decisions made by System 2 because these take effort and are conscious. The systems use different parts of the brain: System 1, for example, uses the amygdala which, amongst many roles, processes emotions, whereas System 2 uses the lateral prefrontal cortex (LPFC). The LPFC is involved in many higher cognitive processes such as working memory, goals, planning and self-control.

## *Short cuts and bias*

Our brains like short cuts. On the whole, short cuts are advantageous to us as they save time and effort – they use the fast System 1 rather than the more reflective System 2. We are confronted with thousands of decisions every day and we cannot think through each one of them. Short cuts mean that we don't have to – on the whole they are a means of guiding us to sensible decisions without having to expend lots of time and energy. But short cuts and bias can be dangerous and lead us into flawed decision-making. They are particularly pernicious because we are unaware of them – we view the world through them and we do not realize that they are colouring and distorting that view. They come into play in every conversation, every meeting, every decision we make. Some biases are short cuts and are about expedience, others are about protecting our egos, and some are based on our values. I used to use a simple fairy story in workshops that was based on a king, a queen, her lover and the castle guard who killed the queen on the instruction of the king because of a rule she knowingly broke. The team had to agree who was responsible for the queen's death. No team ever agreed and discussions became very heated over what was, on one level, just a made-up story. Team members thought others were stupid, uncaring, prejudiced, or sexist. What the story brought out was different values and biases and how hard it is to reach agreement when we view the world through our different biases. Personality, experience, current context and emotions all affect how we see the world. So, if you think back to a time when a colleague interpreted a situation in a very different way from you and you just could not understand how they could read the situation, this might be the explanation. Another of our biases is that we think other people see the world as we do, but they don't.

## Short cuts

Professor Robert Cialdini, in his book *Influence: Science and Practice* (2001), has identified six short cuts that our brains typically take because, on the whole, they are a useful guide as to what is a good decision.

Based on research by Professor Robert Cialdini of the University of Arizona.

1 **Reciprocation**. We have a strong sense that if someone does us a favour, we should repay that favour. If someone sends us a Christmas card, we feel we should send them one. If someone invites us round to their house, we feel we should invite them round to our home. But if you feel someone owes you, act soon – the sense of need to reciprocate declines sooner in the recipient.

2 **Commitment and consistency**. We are more likely to keep to our commitments if we have put them in writing or if we have made those

commitments in front of other people: we want to be seen to keep our promises. We want to fit the image that others have of us and fit the image that we have projected of ourselves.

3  **Social proof** or 'people who are like me'. This influencing principle is about the fact that we often decide what is 'correct' behaviour based on what other people do, particularly people who are similar to us. Professor Cialdini conducted an experiment in a hotel. He and his team created different signs. Sign 1 was the typical sign we see, asking people to help protect the environment by reusing their towels. Sign 2 said that most other guests who had stayed in the hotel had reused their towels – this led to 26 per cent more people reusing their towels than Sign 1. Sign 3 stated that the majority of people who had stayed *in this room* had reused their towels, and this increased reuse by 33 per cent. There are lots of examples of this being used by governments: messages such as '95 per cent of people get their tax return in on time' encourage us to do the same. Messages in doctors' surgeries complaining about the number of people failing to turn up to their appointments lead to more missed appointments: we absorb the message – other people don't turn up, that's the social norm, so it doesn't matter if I don't turn up either.

4  **Liking and rapport**. The more we like people, the more likely we are to say 'yes' to them. Another point to note here is that people are more likely to respond if they receive a *personalized* request.

5  **Scarcity**. Nearly all of us are susceptible to the scarcity principle: opportunities or objects feel more valuable to us the less they are available. Sales people frequently use this as a tactic.

6  **Expert and authority**. If someone is perceived to be an expert we are more likely to listen to that person's opinion. If someone has authority over us we are more likely to comply.

Governments and other institutions are already using these short cuts to influence our behaviour – encouraging us to keep our doctors' appointments, use less energy, recycle more. These tactics can clearly be used as a force for good to encourage us to do things that will benefit both us and society, but there are ethical issues. These 'nudge' tactics exploit the short cuts our brains are prone to taking and that are often taken without our being aware of them. These short cuts need to be used with everyone's best interests at heart. Robert Cialdini argues that it is important that we all become aware of these short cuts so that we are aware of how others might be applying these techniques to us.

## Social conformity

Philip Zimbardo, Professor at Stanford University, has filmed experiments (*The Human Zoo*, 2000) that should act as a warning to us all. One example: a person is recruited thinking that they are about to participate in some market research. The participant sits in a room with 10 or so other people whom they think are other research participants. All of them have been asked to complete a questionnaire: as they do so, smoke begins to seep under a door – a serious amount of smoke that would indicate there's a significant fire just outside the door. In the film, we see the participant look at the others to see how they are responding. However, the rest of the group have been briefed beforehand that this is just an experiment and they should keep on filling in the form, unperturbed by the smoke. The participant takes the lead from the others and sits there too – on average for 13 minutes while smoke continues to fill the room. This is social conformity at work. This happened time and time again. Zimbardo also conducted the same experiment, but this time the research participant is alone in the room. As the smoke appears, the lone participant makes a sensible and hasty exit.

Social conformity occurs because of our automatic response to an ambiguous social situation. We look to others to guide us because we don't know what to do. We want to fit in. This is a useful thought to keep in mind when planning change – employees will be influenced by their co-workers and, as we saw from Cialdini's work, are more likely to follow a behaviour if they think it is the 'norm'.

This tendency to conform is so powerful that it undermines our own judgement about what is the best course of action. When we are alone we take responsibility and act appropriately. Zimbardo's work raises issues that organizations need to be aware of: the extent to which employees will defer to the group and not 'call out' issues of, for example, bullying, dangerous practices at work, or fraud. The first step in dealing with this is awareness: if we know we are prone to this short cut, and have it at the back of our minds in the workplace, then we are better placed to prevent it from influencing our decisions.

## Bias

Over 150 biases have been identified. We will take a look at a few that are particularly relevant to organizations, but first some key points about bias.

## We're all biased

The first, and uncomfortable, point is that we are all biased. As we have seen already in this chapter, our brains take short cuts and often we are not even aware that they are doing this. Bias is automatic in our brains. This means that there is no need to feel guilty, but nor should we be complacent. After all, we are all both perpetrators and victims of bias. The first step is for people to accept that they are biased. The second step is for organizations to make people aware of their biases and the negative impact on individuals, decision-making, and the success of the organization. The third step is to put plans in place to mitigate biases. We will come back to this in the *What can we do?* section.

No one likes to be wrong and no one likes to have their biases pointed out to them. It's an uncomfortable word. Being right, or at least thinking that we are right, is rewarding to the brain and activates the reward network. One of the reasons that change is so hard is that we have to admit that things are not quite right. Anyone who has gone through organizational change will have witnessed this at some stage. The very fact of asking people to change the way they do things might feel like criticism of their current ways of working: something is not good enough and needs to be better. This is part of what leads to resistance (we will come back to this in Chapter 8 on communication).

## Leaders are more prone to bias

Professor Susan Fiske of Princeton University has conducted many experiments into bias and stereotyping. Fiske's research shows that people with power tend to pay less attention to others and tend to individuate less, so leaders in organizations are more likely to stereotype others and 'put them in boxes'. Fiske suggests several reasons as to why this might be: the first is that powerful people don't need to pay so much attention to others, as powerful people have more control over their own outcomes. If you are now in a role with some power and influence, think back to when you first started working, when you were at the base of the hierarchy. At that early stage in our careers we are much more dependent on others and what they think of us: we are likely to pay more attention to what works for them and what doesn't. To progress, we need to catch the eye of those who make decisions about promotion and we need to understand what they favour. As Fiske says, those with less power will pay attention to more senior people because they want to be better able to predict and have some control over what might happen. As we saw in Chapter 2, our brains want to predict because prediction leads to better chances of thriving. Fiske also suggests that leaders might be more prone to bias because they have more demands on their time and more people competing for their attention, which allows them less space to think about each person as an individual. Powerful people might stereotype people less if they paid more attention to them. There is another issue: some people enjoy having power. Those leaders who like power will also have less motivation to consider each employee as a person (Fiske, 1993). Stereotyping others is a form of control over them: putting them into a 'box' where we have certain expectations about their abilities and how they will behave. No one wants to be stereotyped.

It is not just the fact of being a leader that makes people prone to bias: research conducted by Galinsky *et al* (2006) shows that just putting people into a position where they feel they have power changes the way they think about others very quickly. In four experiments, Galinsky *et al* primed one group of people by asking them to write down a memory where they had power over other people, and another group were asked to write about a time when others had power over them:

- In Experiment 1, both groups were then asked to write the letter E on their forehead. People who 'had power' were almost three times more likely to draw the letter orientated so it would be readable by themselves rather than readable by other people.

- In Experiments 2a and 2b, people with power were more likely to assume other people had the same information that they had and therefore would have the same understanding of an e-mail. They forgot that others had not had the same access to information. How many leaders in organizations have made this same mistake?

- In Experiment 4, people with power were less accurate than the control group at judging emotional expressions, when shown faces that expressed happiness, fear, anger or sadness.

As Galinsky *et al* state, 'Across these studies, power was associated with a reduced tendency to comprehend how other people see the world, think about the world, and feel about the world': a useful and sobering message to anyone with power in an organization.

### Intelligence does not reduce bias

Leaders might argue that they are highly intelligent and would be less prone to bias or at least aware of these tendencies. However, if anything, higher cognitive ability is associated with a greater inability to see one's own bias (West *et al*, 2012). Intelligent people are more likely to come up with all the arguments as to why they are right to hold their current views. People with less confidence in their intelligence are more likely to accept that they might have made a mistake.

### Unconscious bias training only takes us so far

Many organizations have introduced unconscious bias training and, done well, this is a useful first step. But the problem with a training programme is that the next time we are tired, feeling stressed or overwhelmed with information, System 1 and automatic thinking will kick in and we will be prone to our biases again.

### Ingroup and outgroup bias

We explored ingroups and outgroups in Chapter 5 on the social brain and, in particular, looked at Baumeister's work on the negative impact of social rejection on our ability to think. Put simply, ingroups are the people with whom we feel safe and who we feel are 'on our side'. People who are in our outgroup are people whom we don't warm to in the same way or feel are 'not one of us' or whom we fear might pose a threat.

### Outgroups

Within fractions of a second of meeting people, our brains automatically categorize them – ingroup or outgroup (and in some cases we have an indeterminate group where we are not quite sure what to make of the person). This categorization is unavoidable and influences our judgements and decisions about others.

Outgroup faces activate the amygdala, the part of the brain that activates our fear response (Hart *et al*, 2000). We are less empathetic and do not try as hard to understand people from an outgroup. This has worrying implications for all of us as we are all both perpetrators and victims. Neuroscience shows that diversity is good for organizations – it is good to have people who can look at issues from a different perspective. But at an organizational level, these automatic feelings and judgements will harm diversity and cross-cultural working. Linville and Jones (1980) showed that people tend to process information about outgroup members in extreme, black and white, simplistic ways (hence prejudice), whereas we process ideas about people who are like us in more nuanced and complex ways. The mere existence of a social bond of some sort leads to more complex processing of thoughts about that person. We process thoughts about people we are close to in the same way as thoughts about ourselves.

### Reducing outgroup bias

Many leaders and managers are concerned about improving relationships between groups within organizations. The cause of much stress and tension in organizations is due to how people feel about other groups of colleagues. Collaboration is essential for success. Face-to-face contact can help to reduce feelings of outgroup. Wilder and Thompson (1980) showed that people form more positive views towards people in their outgroups if they spend time with them. Their research shows that a second meeting with the outgroup members works better than just one meeting. However, the liking only increased for the individuals they met, not for the wider outgroup.

### Ingroups and teams

The terms ingroup and outgroup were made popular by Henri Tajfel. Tajfel *et al* (1970) showed that assigning participants to groups on an arbitrary basis – for example, whether you have a tendency to overcount or undercount a large number of dots on an image, or whether you prefer paintings by one painter you haven't heard of as opposed to another – was enough to lead people to favour those in their group over the other group when asked to share out rewards. When the participants had a choice as to whether to maximize rewards for the whole group or just for their own group, they chose the latter. Tajfel's work shows how quickly we make decisions that discriminate against outgroups, even if these are people we know well. Organizational change can quickly change who is in an outgroup. Working on an integration in the United States, I saw factions and loyalties shift quickly between those who had been offered a role in the newly-integrated organization and those who were still searching for a role. While I was working in the United States, I also saw how quickly the relationship with our UK-based colleagues deteriorated and these were colleagues and friends we had worked with for many years. Misunderstandings arose and people were noticeably less forgiving of each other's mistakes. Looking at Tajfel's and others' research from a more positive perspective, it also shows how quickly we build loyalty and allegiance to those who are in our ingroup.

### Diverse teams – improved critical thinking

Diverse teams can feel less comfortable to work in but this lack of comfort can have a positive impact on the quality of our thinking and our decision-making. Research conducted by Levine *et al* in 2014 tested the performance of traders who were doing business with people either from the same ethnic background or from different ethnic backgrounds. The researchers found that those people who traded with people from the same ethnic background were overly trusting, made more errors and created a price bubble. In a diverse market, traders were more likely to examine and analyse others' behaviour and were less likely to think others' decisions were sensible. They were more likely to consider information with greater care and depth. In this case, ethnic diversity was beneficial not because people brought different views, but because the very presence of someone from a different background improved people's thinking. Working in a diverse group interrupts our brain's shortcut of 'social proof', where we tend to unthinkingly behave as others do. Working with people who are different from us disrupts some of our assumptions. It reduces 'group think' and sharpens up our thinking. Research by Sommers (2006) shows that ethnically diverse juries consider a wider range of views, discuss for longer and make fewer inaccurate statements than homogeneous juries. Diversity is not just good for those who are typically in the minorities – it improves the depth of thinking of all of us, benefiting everyone, minorities and majority alike. An important question for us all: are we working in teams of people who are all a bit like us? If so, our thinking and decision-making might well be lazy and flawed.

## Other forms of bias

The biases above are largely about social biases or biases in relation to other people. There are many different types of bias and here are a few others that organizations need to be aware of during change.

### Confirmation bias (also known as myside bias)

We like to be right and feeling right is rewarding to the brain. So, it is perhaps not surprising that a bias we are prone to is confirmation bias: focusing on information that supports what we already believe and paying less attention to what doesn't support our views. It feels good to have our views endorsed. This can lead to deeply flawed decision-making and even outright danger. Pilots are taught that they must believe their instruments rather than what they think might be happening. The most blatant form of confirmation bias that I have witnessed was feeding back cultural survey results to a CEO. We had been told that the CEO did not have much time, and so we kept the feedback to two pages – the good news and the bad news. The CEO listened to the good news and then told us he had not got enough time to hear the bad news and promptly left the room. At least this was overt conscious confirmation bias – typically it is more insidious than this.

### Sunk-cost bias

We touched on this in Chapter 6: this is the bias that means that it is hard for us to give up on something in which we have already invested a lot of time, effort and resources. An example of the sunk-cost bias is Concorde, in which both the French and UK governments continued to invest even after it became clear that it would never become profitable. This is a large-scale example but it also affects us in smaller ways: not pulling out of investments soon enough, continuing with projects that are not delivering, continuing with employees who are constantly underperforming, and – even in our personal lives – sitting all the way through a film that we are not enjoying. Writing off time and money is painful for organizations but change can be a good opportunity to question the value of projects, practices and processes.

## Temporal discounting

The reward processes in the striatum tend to discount the value of future rewards. The question 'would you prefer £10 right now or £15 in six months' time' tests this bias. Pressure is put on leaders to deliver results now, rather than build for the future. They are judged on relatively short-term results. This same bias leads us to invest less in provisions for the future – pensions, for example – than we should. In change programmes, it means that we need to communicate not just about the benefits that we will see in years to come but also about some of the 'quick wins' and more immediate benefits that the change programme will bring.

## Projection bias

We tend to assume that people see the world the way we do and think like us. This is one of the biases that the CEO I mentioned in the opening section of this chapter fell into: assuming that his fellow directors would see the need for change as he did and would be fully supportive. His position of power might also mean that, as Galinsky's research suggests, he had a reduced ability to see how others see, think and feel.

## Anchoring effect

This bias occurs when the first piece of information we are given has a significant influence on how we perceive subsequent information. Salespeople use this tactic, asking for a high price and then lowering it to one that is still high but seems less so when compared with the first price. Organizations sometimes manipulate their external communication using this effect – putting out bad news in advance so that the not-quite-so-bad actual news does not look quite as bleak. Organizations sometimes use this technique internally – asking for 30 per cent cuts across all departments, whereas in reality 20 per cent will be acceptable. However, this approach will likely cause a greater threat response in the short term and cynicism and mistrust in the longer term.

### Illusion of control

We overestimate our influence over external events. In his book, *Thinking Fast and Slow* (2012), Kahneman describes analysing the results of a group of investment advisers – his results showed that the advisers' advice was as useful as if they had rolled dice. What does this bias mean for organizations going through change? We need to think about what we are really rewarding in organizations – skill or luck. It also means that before embarking on change, leaders need to be honest with themselves about what they can influence and what might be out of their control. Being realistic and honest might mean that organizations do not embark on change programmes that are too risky. It also means that they need to ask searching questions of suppliers and whether they can really deliver, for example, the IT programme they say they can.

### Loss aversion

In *Thinking Fast and Slow* (2012), Kahneman provides examples of various forms of loss aversion: a toss of a coin – heads you win $150, tails you lose $100. Is this a gamble in which you'd feel comfortable to participate? Most people would not. The pain of the potential loss outweighs the pleasure of the potential gain. This bias has a particularly important bearing on change: in many change programmes people stand to lose something – for example, shared services can mean a loss of autonomy but access to better resources. The pain of the loss outweighs the pleasure of the gain. This might not be rational but, as we know, the threat response is stronger than the reward response. We need to deal sensitively with employees who have to give up something during a change programme and not be surprised by the pain they feel. To our brains, losses loom larger than gains.

Kahneman provides another example of loss aversion influenced by the way in which decisions are framed. People were told that a disease was about to break out and it was expected to kill 600 people. When presented with an option that will definitely save 200 lives (but will risk 400 dying), people will choose to save 200 lives. When presented with the same option but negatively framed (400 will die), people are not happy to make that choice. The second is framed as a loss which people are averse to (Tversky and Kahneman, 1981).

### Beware the lure of certainty

As mentioned in earlier chapters, our brains crave certainty and find living in uncertainty uncomfortable and distracting. When making decisions we need to beware the lure of certainty: making a decision too quickly because the closure that it offers is attractive and is more comfortable than living a little longer in uncertainty. We need to keep Ratcliff's diffusion decision model in mind – waiting to reach the outer thresholds might be uncomfortable and leave us in limbo for longer, but it might lead to a better decision.

**Language and memory**

As every courtroom barrister will know, the language we choose to use can influence people's memories. This was illustrated by Elizabeth Loftus and John Palmer (1974) when they showed films of traffic accidents to participants. Participants were asked at what speed they thought the vehicles were travelling when they 'smashed/collided/bumped/hit/contacted'. The use of the word 'smashed' elicited higher estimates of the speed. Not only that, but one week later, those participants who had heard the 'smashed' verb were more likely to answer positively when asked if they had seen broken glass in the film, even though there was no broken glass. This research shows the impact of language on how we reconstruct memories. There are lessons here, too, for organizations and employees and how they talk about the past, and the need to choose our words responsibly.

## Mindfulness and reduction in bias

We saw in Chapter 6 that mindfulness can improve decision-making through recognition and reduction of cognitive biases (Hafenbrack *et al*, 2014). Research by Kirk *et al* (2011) demonstrates that people with mindfulness training can separate their emotional reactions from their behaviour while playing the ultimatum game. People who practise mindfulness were twice as likely to accept the free money rather than take no money at all: they are able to detach themselves from the emotional reaction of feeling that the offer is unfair. When considering unfair offers, people who practise mindfulness were seen in fMRI scanners to have reduced activity in the anterior insula and to use a different brain network (Kirk *et al*, 2011). Mindfulness helps people to become more aware of their thinking patterns and so provides greater awareness when we are falling into known biases.

These are just a few of the many biases that affect our decision-making. The challenge for us is that there are so many that we cannot be constantly aware of them: by their very nature we are blind to them and we cannot rely on our self-control to manage them. We will look at some of the steps organizations can take to reduce the negative impact of bias in the next section. Neuroscience is beginning to reveal some insights.

# What can we do? Solutions and examples of what other leaders have done

## *Decision-making*

This section takes a look at the steps organizations can take in light of what neuroscience teaches us about decision-making and bias.

### Plan decision-making meetings

Decision-making depletes so organizations need to:

- Hold decision-making meetings when people have energy (earlier in the working day rather than later).
- Consider how many 'big' decisions should be on the agenda.
- Think about how much data people can take in.
- Supply refreshments to maintain steady glucose levels (remember the judges in Chapter 6).
- Think what kind of solutions they are after – ones that can be reached through analysis or ones that need insight? If the latter, sitting in a meeting will not be the best place to reach insight.
- Reflect on whether the decision should be taken straight away – might sleeping on the idea help? There's more on the brain's need for sleep in Chapter 11.
- Make sure employees take some time out – go for a walk, let minds wander for a while, then come back to the decision.
- Use diffusion models for decision-making – think about the trade-off speed/accuracy.
- Be aware of Systems 1 and 2 – are the participants giving too much sway to System 1?
- Mitigate biases – see below.

## *Overcoming bias*

Overcoming biases is difficult. The desire not to be biased is not enough. We are blind to our own biases and none of us likes to feel that we are biased, mistaken or just plain wrong. Biases operate without us even being aware of them so we can't find a cognitive way to deal with them. Over 150 biases have been identified – too many for us to remember. What's more, even when we are made more aware of them, the next time we are tired, short of time or feeling stressed, we will slip back into being biased. There is no easy solution to dealing with biases but there are some practices we can put into place that will help. A few key principles:

- Build awareness of the fact that our brains lead us to be biased: being aware and vigilant is the first step. People do not need to feel guilty, but nor should they be complacent.

- Biases are largely unconscious and we can see others' biases better than we can see our own, so get people to help and challenge each other.

- Be open to challenge; be curious about different views.

- Do not rely on self-control, remembering not to be biased, etc: it doesn't work.

- Put plans in place that prevent biases from colouring decisions: reminders at the start of meetings; guidelines as to how decisions should be made.

- It is important to be able to see the limitations of your own thinking, so set some time aside to identify your own mistakes.

- Bias creeps in when goals are unclear, so set clear and fair procedures to guide behaviour.

- Stick to the evidence.

This reminds me of working with an investment bank. They had a head of diversity and were aware, in theory, of the need for diversity. Yet, speaking to one leader just before he headed off to a business school to find new recruits, I asked what criteria he would use to select people. He said he just knew when he met people whether they felt like 'one of us'. People need clear goals, procedures and plans but to remove bias, businesses do need to make sure these are being followed, even by the mavericks.

## Mindfulness

The research of Hafenbrack and Kirk and colleagues is beginning to build the case for how mindfulness can protect us from bias. See Chapter 6 for suggestions on how to get started with mindfulness.

## Physiological needs

We know that glucose levels have a major impact on our decision-making: we should check that people have eaten before decision-making meetings.

But what about sleep? There's a culture in some professional services firms that rewards people who work very long hours. But can these people really do their best thinking and should they be making decisions that have significant consequences? Sleep deprivation is like being drunk (Dawson and Reid, 1997). These firms would not allow people to be intoxicated and making decisions or in charge of machinery, so why let people work such extensive hours? Sleep supports memory, insight, emotional regulation, and enables us to see connections and patterns. People are not machines: if we want to work at our best, we need to look after our brains.

## Devil's advocate

Asking one person in the group to play devil's advocate and to challenge the assumptions that are being made is a useful way to surface biases and lazy thinking. Warren Buffet, the very successful investor, says he sometimes does this: he invites someone whom he knows has very different views to put forward different ideas and to tell him why he is wrong.

In some of Shakespeare's plays, the role of the court jester or fool was often to say what others dared not say. The fool is the one person who can speak the truth and challenge the king and act as his conscience. Organizations need to find people who can challenge and question. This role of acting as the organizational conscience is one that can be played by Communications – especially communicators who are responsible for employee communication. They are often well placed to act as the representatives of employees, and to feed back to leaders what people think and feel about the change programme. New joiners and outsiders, such as consultants, have the benefit of being able to look at the organization more objectively and can ask the 'dumb question' that people inside the organization feel they can no longer ask without risking looking stupid.

## Prime ourselves not to be biased

We cannot rely on our self-control to stop us falling prey to our biases. We need to put plans in place. After unconscious bias training, one manager in a government department said that the team now states at the start of the meeting how they will make decisions that are as free of bias as possible, eg 'We will be fair'. The act of stating this raises people's awareness of the principle and makes it easier for members of the group to challenge during the meeting whether they are indeed being fair.

## Be aware of our values

We see and judge the world through our values that are deep-rooted and often unconscious. Self-awareness is key: if we know our own values, we will have some insight into how we see the world and how we judge others. Here's a simple exercise to help people discover their values and their potential biases. People might think their values should be obvious to them but this exercise can reveal some unexpected insights and clarity. Asking people to do this in pairs and to interview each other works well.

1  List three things you value with regard to each of the following categories:
   - friendship (eg what three things do you value most in friendships?);
   - yourself;
   - life;
   - family;
   - work;
   - leisure time/hobbies.

2  List them all and see which values come up in more than one category.

3  Ask which is the most important value. Why is it so important? What does it give you?

4  Which is the second most important? Why? What does it give you?

This is an exercise that some coaches use with clients to help them understand their motivations and to understand what fulfilment looks like. For our purposes here, it is to help people be clear about their values and therefore provide more clarity on how they view the world so that they can be aware of how they might be judging others.

## Remove defensiveness

We are more prone to some biases when we are feeling negative or in a threat state. So one useful tactic is to put people into a positive 'toward' state that activates the ventral striatum – part of the reward network – before decision-making meetings. This will make them less vulnerable to those biases that are about protecting ego.

## Be aware of ingroup/outgroup bias

One of the most frequent insights that managers have once they are aware of ingroup/outgroup bias is the impact it can have on decisions about others. One group of leaders in financial services looked at the performance ratings and realized that 'ingroup bias' had affected their judgement. Another recognized that they had been blaming their inconsistent Net Promoter Score on people whom they did not know so well, whereas in fact the responsibility lay with someone in their ingroup – they realized that the 'too comfortable' relationship had undermined their ability to be objective and fair. Other leaders and managers could do well to follow suit: to review their decisions, especially those that are about people's performance, and check for ingroup and outgroup bias.

## Build ingroups

Chapter 5 on the social brain provided plenty of information on why people want and need to feel part of an ingroup. Diversity is a good thing: neuroscience research suggests that people from different world cultures see the world differently, and this range of views brings breadth of knowledge and experience to the organization. As we saw earlier, diverse groups also tend to have less lazy and better critical thinking. To reduce outgroup bias, there are two practices that work well:

1 Create superordinate goals: shared goals that bind people together.
2 Familiarity breeds empathy: face-to-face communication makes a difference.

Creating shared goals is a great way to bring people together: it enables people to see what they have in common rather than what separates them. It is a simple change that very quickly can make an immediate difference to how people feel about each other and on how people work together.

The second point, face-to-face communication, can be difficult in geographically-dispersed organizations, but time and time again people report how differently they feel about their colleagues once they have met them face-to-face. On a global project for an investment bank, our one key learning at the end of the project was that we should have met sooner face-to-face. Before we met, there were doubts about some of our colleagues. Once we had met, we were noticeably more empathetic. The more familiar people are to us, the more the brain thinks of them in the same way as we think of ourselves.

## Confirmation bias – remain open to challenge

One of the most important things we can do to protect ourselves from our biases is to be open to challenge. The world of aviation has some useful practices to share. The Captain has responsibility for the safety of the plane.

Relationships between the Captain and the First Officer used to be very hierarchical and were referred to as the 'Captain God complex'. After some accidents, the black box recordings revealed that the First Officer had seen there was a problem and that the Captain was making a mistake, but lacked the language or confidence to challenge. To reduce the risk of accidents, flight crews are now trained very differently so that the Captain is open to challenge and the First Officer feels confident to do this. Captains remaining open to challenge is crucial for the safety of planes, and leaders being open to other ideas is essential for the success of organizations.

The attitude of the leader in meetings is crucial. I observed one executive team meeting in which a person who had a different view from the CEO was ridiculed in front of the team. Not surprisingly, after that everyone stayed very quiet. On the other hand, I participated in a very well-facilitated leadership meeting in which each person was encouraged to give their view, was listened to, and then the CEO made a decision based on all that he had heard.

A lot of pressure is placed on politicians and leaders to be consistent and not to change their minds. Changing their minds is seen as a weakness and something that the media will not forgive. 'The lady is not for turning', Margaret Thatcher proudly announced in 1980. People want predictability and certainty from their leaders, but this pressure never to change their minds, never to admit that they might have got something wrong, is a double-edged sword. It can be a sign of strength but it can also lead to weakness if leaders fail to heed the warnings that they might need to think again.

# Summary of key points from this chapter

## *The science*

- Change often requires decisions to be taken swiftly, but it also means that we are making decisions in an environment of uncertainty, which means our brains might be in a 'threat' state where we see the world as more negative than it really is.

- Making decisions of all types depletes our mental energy and this affects the quality of our subsequent decision-making.

- Glucose has a significant impact on our ability to focus and on our decision-making ability.

- Emotions play an important part in decision-making: the ultimatum game illustrates this.

- Without emotions we find it very hard to make the simplest of decisions. Decisions based just on reason are not always sound and not all decisions influenced by emotion are flawed.

- The diffusion decision model is a useful mental model and acts as a reminder that there is always a trade-off between speed and accuracy.

- We have two thinking systems: one fast, automatic and reflexive that we are largely unaware of (System 1); the other slow, reflective and effortful (System 2). Our brains tend to use System 1.

- Our brains use short cuts and, on the whole, these are helpful as they save time and effort, but they can lead to flawed thinking.

- Research into social conformity shows how influenced we are in our decisions by what our peers do. This is a useful point to note for organizations trying to change the behaviour of employees.

- We think we see the world objectively but we are looking at it through a lens that is shaped by our past experience, our current context, our expectations, our emotions, our personality and our biases: we are not as objective or rational as we think we are.

- Looking at the world through our biases means that it can be very hard for us to understand why others don't see the same situation exactly as we see it.

- We are all biased. Our brains make automatic judgements within fractions of a second.

- We are blind to our biases.

- People with power are more prone to bias, and intelligence does not help – in fact, it might make bias even more ingrained.

- Unconscious bias training only goes so far: we very quickly slip back into our biases.

- Research shows diverse groups can perform better than homogeneous groups because of better critical thinking.

- There are over 150 identified forms of bias – some of these we have developed for expediency, some for self-protection.

- We process thoughts about people in our outgroups differently from those in our ingroups. We think about people in our ingroups in a more complex and nuanced way. Outgroup faces activate the amygdala.

- We form ingroups and outgroups very quickly and discriminate immediately.

- Mindfulness has been shown to protect us from bias.

## *What can we do?*

- Think about the environment and conditions that enable people to be at their best for decision-making: time of day, speed vs accuracy, number of decisions on the agenda.
- Pay attention to physiological needs that have an impact on our ability to make good decisions – glucose levels, water, sleep.
- Put practices in place that reduce bias:
  - Build acceptance that we are all biased – it's what our brains do.
  - Don't let people rely on self-monitoring – we can't do it.
  - Unconscious bias training has its limits: set up systems that reduce the influence of bias.
  - Set clear goals and guidelines.
- Remain open to challenge and different views.
- Ask someone to play devil's advocate in meetings to question the thinking.
- Prime people not to be biased.
- Identify your own values and potential biases.
- Remove defensiveness.
- Be aware of your emotions but keep to the evidence.
- Check decisions you have made about people against ingroup/outgroup bias.
- To form ingroups:
  - Set shared goals.
  - Create opportunities for face-to-face communication.
- Research studies show that practising mindfulness helps to reduce bias.

# Reflections and planning

We know that there is pressure during change programmes to make decisions swiftly and we know that our brains have a limited capacity for making decisions, so:

- How will you make sure that you are finding the right balance between speed and accuracy in decision-making?

- Have you reviewed your agendas: do you think the number of decisions on the agenda means that you can give the right level of thought to each to ensure they will be good decisions?
- Is there a pressure to make decisions straight away? Is there a need for more time to reflect?
- How are the meetings structured: do the meetings allow for physiological needs (sleep, glucose levels, etc) that can have such a strong impact on decisions?
- What about decision-making meetings about people – promotions, appraisals, bonuses: how will you ensure each person is getting fair consideration?
- How diverse is the group making the decision?

We are all prone to bias and we are blind to our biases, so:

- How will you build knowledge and acceptance in the organization that we are all biased – people with power more so than others?
- How will you promote the benefits of reducing bias?
- What kinds of bias have you seen in operation in your organization?
- What systems do you currently have in place to reduce bias? Do you need to put more systems into place? If so, how do you do this so that people see the benefits and don't feel it's just bureaucracy? How do you involve people in developing these guidelines?
- How aware are people in your organization of their personal values and how these influence how they perceive the world?
- How do leaders in your organization challenge their own thinking?
- How open are leaders to the different views of others?
- What examples are there in your organization of people being able to challenge constructively?
- Social conformity means that we are less likely to 'call out' bad and dangerous practices, eg bullying, fraud, etc unless we see others do it first. What will you do to make people aware of this tendency and enable them to speak out?
- How acceptable is it for leaders in your organization to change their mind?
- How can you ensure others put systems into place to reduce bias locally?

We very quickly discriminate against outgroups and in favour of ingroups, but we know that diversity improves decision-making, so:

- Where does outgroup bias have a negative impact in your organization? What examples have you seen? Are leaders aware of these examples?

- How can you reduce outgroup bias?

- What opportunities are there to remind different groups of what they have in common?

- Familiarity increases positive feelings towards others – how can you create more opportunities for different groups to meet?

# References and further reading

Baumeister, R F and Tierney, J (2011) *Willpower: Why Self-Control is the Secret to Success*, Penguin, London

Camerer, C and Thaler, R H (1995) Anomalies: ultimatums, dictators and manners, *The Journal of Economic Perspectives*, **9** (2), pp 209–19

Cialdini, R B (2001) *Influence: Science and Practice*, Allyn and Bacon, Needham Heights MA

Dawson, D and Reid, K (1997) Fatigue, alcohol and performance impairment, *Nature*, **388** (6639), p 235

Fiske, S T (1993) Controlling other people: the impact of power on stereotyping, *American Psychologist*, **48** (6), pp 621–28

Frederick, S (2005) Cognitive reflection and decision making, *Journal of Economic Perspectives*, **19** (4), pp 25–42

Galinsky, A D *et al* (2006) Power and perspectives not taken, *Psychological Science*, **17** (12), pp 1068–74

Hafenbrack, A C, Kinias, Z and Barsade, S G (2014) Debiasing the mind through meditation: mindfulness and the sunk-cost bias, *Psychological Science*, **25**, pp 369–76

Hart, A J *et al* (2000) Differential response in the human amygdala to racial outgroup vs ingroup face stimuli, *NeuroReport*, **11** (11), pp 2351–55

Kahneman, D (2012) *Thinking Fast and Slow*, Penguin Books, London

Kirk, U, Downar, J and Montague, P R (2011) Interoception drives increased rational decision-making in meditators playing the ultimatum game, *Frontiers in Neuroscience*, **5**, p 49

Levine, S S, Apfelbaum, E P, Bernard, M, Bartelt, V L, Zajac, E J and Stark, D (2014) Ethnic diversity deflates price bubbles, *PNAS*, **111** (52), pp 18524–29

Linville, P W and Jones, E E (1980) Polarized appraisals of out-group members, *Journal of Personality and Social Psychology*, **38** (5), pp 689–703

Loftus, E F and Palmer, J C (1974) Reconstruction of automobile destruction: an example of the interaction between language and memory, *Journal of Verbal Learning and Verbal Behaviour*, **13** (5), pp 585–89

Ratcliff, R and McKoon, G (2008) The diffusion decision model: theory and data for two-choice decision tasks, *Neural Computation*, **20** (4), pp 873–922

Sommers, S R (2006) On racial diversity and group decision-making: identifying multiple effects of racial composition on jury deliberations, *Journal of Personality and Social Psychology*, **90** (4), pp 597–612

Tajfel, H *et al* (1970) Experiments in intergroup discrimination, *Scientific American*, **223** (5), pp 96–102

Thaler, R H and Sunstein, C R (2009) *Nudge: Improving Decisions about Health, Wealth and Happiness*, Penguin Books

Tversky, A and Kahneman, D (1981) The framing of decisions and the psychology of choice, *Science*, **211** (4481), pp 453–58

West, F W, Meserve, R J and Stanovich, K E (2012) Cognitive sophistication does not attenuate the bias blind spot, *Journal of Personality and Social Psychology*, **103** (3), pp 506–19

Wilder, D A and Thompson, J E (1980) Intergroup contact with independent manipulations on in-group and out-group interaction, *Journal of Personality and Social Psychology*, **38** (4), pp 589–603

Zimbardo, P (2000) *The Human Zoo*, London Weekend Television, UK

# Communication, 08
# involvement
# and the role of
# storytelling

**C**ommunication during change is critical: the change programme might
be full of brilliant ideas, but if employees do not understand it, don't
support it, or feel they are incapable of delivering it, then it won't happen.
Change happens when people start to do things differently and good communication is essential in helping effect this. Communication is always
important to us, and all the more so during change. As we know, our brains
crave information: it all goes back to survival. For our ancestors, the more
information they had, the more likely they were to be able to make good
decisions and survive. Our 21st century brains still have this intense need for
information, and if we don't have it we feel uncomfortable, and so we speculate and fill the gap. This speculation takes up a huge amount of employees'
time and energy. Some leaders and managers see communication as a time-
consuming task that is in addition to the 'real' job. But if they don't communicate well during change, employees will be anxious and distracted and
time will be wasted. As we saw from Galinsky *et al*'s research (2006) in
Chapter 7, people with power tend to assume that others have the same
information that they have. They lose the ability to put themselves in the
shoes of other people and tend to see the change programme just from their
own perspective. A large part of communication during change is to keep
leaders in touch with how employees feel about the changes so that their
communication to employees not only has the right content but has the right
tone – knowing when people need some vision, energy and drive from their
leaders, and knowing when they need empathy and listening.

The initial communication is crucial: shortly after the announcement of a merger in the United States, the new CEO arrived at the headquarters in a limousine, late, and dressed in a suit. His opening words were, 'This is the happiest day of my life'. Employees in the audience were already annoyed at having to wait for a man who arrived dressed very differently from them (so already a strong 'outgroup' feeling about the new CEO) and then were even more angry at what they perceived to be his insensitive comments – some of them felt this might be one of the worst days of their lives as they might lose their jobs. This was a bad start and employees referred to it over and over again amongst themselves, but no one dared to mention it to the CEO. When eventually he was asked about the comment, he said he had spoken with the best of intentions, genuinely feeling that the merger marked a great future for both companies. It took quite a while for him to rebuild those bridges. This start is in sharp contrast to another American CEO arriving at the headquarters of a company after a hostile acquisition. She had asked what might be a way of showing her respect to the acquired company and was told that wearing a jacket in their much-loved corporate colour would go down well. The fact that she had made this effort was appreciated and a first step in mending the relationship.

Change means that there is usually a pressure in organizations to do things very quickly, and communication and getting employees' input can be two of the areas that suffer as a consequence. Leaders feel the pressure to crack on and make change happen, to be seen to be doing something rather than spending time talking with employees. But telling employees they need to change, rather than allowing them the time to reach their own insights, causes a threat or 'away' response in the brain and means that implementing changes can be a lot harder and take a lot longer. Certainty and autonomy are closely linked: when we feel we have control over our work, we have more certainty.

One of the means of enabling employees to reach insights is storytelling. Human beings have an insatiable interest in stories – be these real or fictional, about people in the workplace or people we have never met. Wall paintings in caves that date back around 40,000 years show that our ancestors also enjoyed depicting stories. Much of our time is spent in sharing personal stories, anecdotes and gossip. We go to films, we read, we search on line, we watch television, we listen to the radio, we go to the theatre – and in all these places we are told stories. Our interest in stories starts at a very young age and, for most people, this continues throughout life. Stories in the workplace are growing in popularity but many managers are bemused as to why they are being asked to 'tell stories'. As we'll see later in this chapter, neuroscience provides some explanations as to how and why they work.

# The science

## *Communication*

Our brains are constantly in search of information. Research conducted by Bromberg-Martin and Hikosaka (2009) suggests that not only do our brains want information, but the sooner we get it, the better. The pair worked with macaque monkeys, and even though the monkeys would get later cues about the reward they would get, if they could opt for earlier cues, they would. Bromberg-Martin and Hikosaka found that midbrain (substantia nigra and ventral tegmental area) dopamine-releasing neurons were activated by anticipation of information: in other words, information is rewarding to the brain in the same way as food and water and other 'rewards'. Even when we cannot influence the outcome, we like to get information about upcoming rewards – the promotion we are after, the bonus, exam results that we are feeling confident about – we like to get the information as soon as possible. As we saw in Chapter 3 from Wiggins *et al*'s research (1992) on the Huntington's Disease gene, people prefer negative information to no information. Having bad news meant people were at least able to get on and plan. If leaders are in doubt about time spent on employee communication, remind them that employees' brains are more settled when they have information that will help lead to certainty.

### Grasping at certainty

Uncertainty registers as an error or a gap and something that must be dealt with before we can feel comfortable again. It activates the amygdala (see Chapters 2 and 3) and this makes us anxious. Our brains much prefer certainty. The danger of this can be that, because the brain finds uncertainty and ambiguity uncomfortable, we latch on to what we believe to be the solution too quickly. Thinking that we have found a solution feels good and is rewarding to the brain. As we saw in Chapter 7, leaders going through change might be vulnerable to this – too quickly grasping at what might seem to be 'the answer' when actually they need to live with uncertainty for a little longer.

### Communicating sooner or later – a dilemma

At the start of the 2008 banking crisis, two banking CEOs were faced with the same questions and uncertainties, but took different approaches to communication. One saw that the future was precarious and unpredictable and he decided to work with people in the organization, exploring the potential scenarios, discussing what they knew and what they didn't know. The other took a very different view. Concerned that he might say something one week that would be proved to be wrong the next week, he chose to withdraw and say little. There was no easy answer in this situation: as we know, our brains want certainty, and for banks in 2008 there was very little. Anecdotal reports suggested that employees were happier with the first approach than the latter.

There is a dilemma for organizations going through change as to how much to communicate and when. Communicate too soon and the organization might have to retract or amend some of what it has said. As we saw in Chapter 7, there's an internal and external pressure on us to be right. We put pressure on leaders to be consistent, so communicating when much is still unknown is uncomfortable. However, if leaders wait to communicate until they have more of the detail, there is a communication void. This registers as an error alert in employees' brains and they are distracted and less focused on work until the gap is filled.

## Past experience and expectation influence what we take in

Why is it that people can go to the same meeting and leave having heard different things? Why do leaders think they have communicated a clear set of points, and yet some employees say they have not heard them? Past experience, current context, personality, expectations and biases, all have an impact on how we perceive the world and what we take in. It's one of the reasons why diversity is so useful in organizations – people from different backgrounds hear and see different things and bring a different perspective. But for organizations, when there is a need to communicate consistent messages and get alignment in understanding, this can be a challenge. Our brains constantly subconsciously compare the current situation with past experience. Different past experiences mean that people experience the same information differently – a negative experience of change in the past will mean that those emotions are reawakened, a positive experience of past change will mean that an employee is more likely to look at current changes in a favourable light. In addition, confirmation bias (see Chapter 7) means that we look for information that supports what we already believe and ignore the points that don't fit with our views.

Expectations also have an impact on what we experience. Tetsuo Koyama *et al* (2005) conducted experiments on people that showed that our expectations significantly influence the neural processes that underlie how much pain we actually experience. If people expected less pain, then they reported experiencing less pain and this was backed up by less activation in the brain's pain network.

This explains in part why employees' reactions to change can be so very, and perplexingly, different. What each of us sees as reality is highly subjective. In change, we need to be ready for these very different reactions, and just because someone's take on a situation is different from ours does not necessarily mean they are perverse or stupid. This is one of the reasons why managers play such a key role during change – they are better placed to see what each team member understands and how they are reacting to the changes. Managers can deal with each person individually.

### Setting the right expectations

Think about a time when you didn't get the outcome you wanted – the pay rise you were promised, the exam result you wanted, the recognition for a successful project you deserved. Not reaching your expectations is disappointing and leaves you feeling de-energized and sad. In our brains, unmet expectations lead to a drop in dopamine. It has a negative impact on the PFC and so on our ability to think. Our performance drops and, as we become aware of this decrease, we descend into a vicious cycle. On the other hand, if our expectations are met or even exceeded – sales are even higher than our target – this leads to an increase in feeling good and in dopamine. This in turn enables us to perform better – a virtuous cycle of performance. The implication of this is that we need to be careful in communication about how and where we set employees' expectations. Goals that are overly ambitious and which cannot be achieved will lead to a drop in dopamine and activation of the amygdala. Goals need to be stretching, challenging and obtainable.

### Cognitive dissonance

Our brains constantly compare incoming information with predicted information. If incoming information fits, then all is OK and our brains pay little attention. But if incoming information doesn't fit with our expectations our attention is diverted until we have resolved the issue. Cognitive dissonance is a term that was first used in the 1950s by Festinger. Put simply, it is the uneasy feeling we get when we have conflicting ideas. Change itself produces dissonance because it suggests that the current way of doing things is no longer good enough or quite right. We need to learn new skills or new approaches to how we work. Festinger found that the brain does not like to hold cognitions that are in conflict with one another: dissonance means that one of our ideas is wrong (Festinger *et al*, 1956). As we know from Chapter 7, we hate to be wrong and this leads to a threat state. The brain therefore tries to do one of the following:

1 Remove one of the ideas/disbelieve it.

2 Allow one idea to carry more weight than another.

3 Add or create a new idea or ideas that resolve the conflict.

When we experience cognitive dissonance, we have greater activation in the dorsal Anterior Cingulate Cortex (dACC) and in the insula (van Veen *et al*, 2009) – both are part of the pain network. Conflicting ideas are painful to us. They are also connected to our fight or flight system – this accounts for the visceral feeling we sometimes get when we know we are wrong.

Cognitive dissonance also explains why it is difficult and uncomfortable for us to have contradictory attitudes and actions with regard to ourselves. This is one of the reasons why asking people to make a public statement about a commitment can be a good reinforcement. For example, if a leader has made a public statement about involving employees during change, it is uncomfortable for the leader to go back on those words. People expect the action and the leader needs to do it to avoid cognitive dissonance. Cialdini noted this in his book *Influence* (2001): asking people to make public commitments means they feel more obliged to do it.

The concept of cognitive dissonance brings useful insights to how we communicate change in organizations. On the one hand, cognitive dissonance combined with confirmation bias (see Chapter 7) can keep people connected to the ideas they already have. In this case, leaders need to work hard to enable people to let go of their old ideas. On the other hand cognitive dissonance also means that we will change our ideas and behaviour to fit in with how we see ourselves and how we believe others see us. Cialdini gives examples of this in his book. He cites Henry Kissinger, US diplomat and political scientist, observing how former Egyptian President, Anwar Sadat, was particularly skilled at using this aspect of cognitive dissonance. Sadat primed his negotiating 'opponents' by telling them ahead of discussions that they were known for their cooperativeness and fairness. This put a pressure on them to live up to their reputation. Research by Cioffi and Garner (1996) gives another example of influencing people's behaviour by shaping how they see themselves and how they think others see them. In their research, volunteers who were asked to commit in writing to helping with an AIDS education programme were more likely to do so than those who had volunteered but who had not written this down. Active commitments affect our self-image and this influences our behaviour which further embeds our self-image.

Within organizations, cognitive dissonance provides ideas about how to communicate with people about change, but there are also ethical considerations. Influencing people so that the outcome for all is beneficial would be seen as a constructive act, but manipulating employees without their awareness that this is being done is a dangerous path to tread. Arguably, interest in employee communication grew in the late 20th century because organizations needed to be agile and to adapt quickly. Organizations recognized they could not do this successfully without the support and input of employees. Organizations need to tread a careful line, using communication and influencing techniques for the good of all stakeholders, including employees, and so maintaining an honest and trusting relationship with those employees. This is similar to the dilemma governments face when using 'nudge' tactics (see Chapters 7 and 10) with citizens: the challenge is how to nudge and influence towards a good outcome for all.

Cognitive dissonance also has an impact on communication when leaders say something that is unexpected. The dACC sends an 'error alert' – what was just said does not fit with what we had previously thought. One example of this was a CEO who had been on a campaign to change the way the company negotiated with the unions. Tension was escalating, unions were contemplating industrial action, a lively communications campaign had been launched by both the company and unions. Then suddenly the CEO said the changes were 'just a suggestion'. It was one of those moments when people questioned whether they had heard correctly what he had just said: it did not fit with anything that had been said or done over the previous few weeks. It meant everyone having to readjust how they saw the situation, trying to make sense of this new idea in the context of what had gone before: cognitive dissonance at work. It is important for communications professionals to be aware of this, because if a leader suddenly says something that doesn't fit or is surprising, people will be distracted, trying to make sense of the conflicting ideas.

## Blind spots

There is a limit as to how much our brains can take in. Here are a couple of films – take a look at them right now if you can. The first from Daniel Simons and Christopher Chabris asks us to count how many passes of the ball are made by the team in white:

**https://www.youtube.com/watch?v=vJG698U2Mvo**

and the second, from Richard Wiseman, is The Amazing Colour Changing Card Trick:

**https://www.youtube.com/watch?v=v3iPrBrGSJM**

Without wanting to provide a spoiler, we don't perceive as much as we think we do. Our brains are 'limited capacity machines' and are constrained in how much they can take in. In both these films we are so focused on one task that we miss what else is going on. Change also means that our brains' capacity is being overloaded (see Chapter 3) and since there is a limit to how much we can absorb, we need to plan employee communication even more carefully when going through change and frequently check how much employees have been able to take in.

## Well-intended words

'Can I just have a word with you?'
'Can I give you some feedback?'
'Let me give you some advice...'

Be honest, how do these phrases make you feel? For most of us the feeling is, 'I'd rather you didn't...'. These are phrases that many of us frequently use in the workplace without thinking twice. They might be well-intentioned, but even well-meaning advice can trigger a threat response in our brains, because it suggests that you are not doing things quite right and someone else knows better. People asking if they can give us some feedback fills us slightly with dread: What are they going to say? What have I done wrong? They have noticed, I'm not very good at this. Do they really think they know better than me? Change means that our brains are already on the lookout for threats, and words that cause more uncertainty, or suggest we are about to be criticized, will increase this sense of anxiety. Leaders might not be able to avoid using these phrases, but if they can reassure people quickly (if appropriate), they will get people's attention back faster. After a neuroscience and leadership workshop, one leader, struck by the idea of the 'social brain', put one-to-one meetings in his team's diaries. His intentions were good: he wanted to spend some time with each one. Employees seeing these unexplained meetings suddenly appear in their diaries got nervous – what did he want to tell them? He quickly realized that he needed to reassure them that there was nothing to worry about. This is an example of how easy it is to mean well but trigger a threat response.

## Screens vs face-to-face

Most organizations rely heavily on communicating via screens. Younger employees coming into the workplace are used to communicating this way.

There are conflicting views as to what difference this makes to our brains and there is a need for more evidence. Communicating via devices is almost inevitable in the current workplace and they are a great means to enable employees to share information across locations, especially globally-scattered sites. However, as we saw in Chapter 7, face-to-face communication makes a difference to creating ingroups. It can be more time-consuming than communicating via devices but some face-to-face communication during the change process will be vital for maintaining relationships and trust.

## Warmth leads to trust

For many leaders it's a dilemma: how to get the balance right in their communication between being likable and being respected. Is it better to be loved or feared? Amy Cuddy is a social psychologist who has used experimental methods to investigate how people judge and influence each other. Her 2013 *Harvard Business Review* article with Matthew Kohut and John Neffinger, 'Connect, Then Lead', sets out the two key traits that we notice in leaders: warmth and strength. The article states that many leaders tend to focus on displaying their strength and competence but a growing body of research suggests that the best way to lead and influence is to show warmth first, then competence. Warmth, they say, enables leaders to connect with people and build trust. By projecting warmth through their communication and behaviour, leaders are more likely to be listened to by employees. Trust in turn 'provides the opportunity to change people's attitudes and beliefs, not just their outward behavior'. So for sustained commitment to change, leaders need to convey empathy. Cuddy *et al*'s research fits with what we know about the social brain and about ingroups and outgroups (see Chapters 5 and 7).

## Listening

We all know how good it feels when someone is really listening to us and they are undistracted, interested and empathetic. I remember once working with a senior partner in an accountancy firm who, while I talked to him, would move around his office, looking out the window, fiddling with paperwork. He seemed distracted. When I challenged him about this, he proved that he had heard every word I had said – it was just that I had found it hard to speak. It brought home the point that we listen not just to help ourselves but to help the speaker. Often employees say that one of the most important things for them when going through change is having someone just listen. Behavioural science backs this up. As we saw in Chapter 5 from Baumeister and Leary (1995), positive social interactions are rewarding, and from Lieberman and Eisenberger (2009) that they activate the reward network in the brain. This matters because when we are in a 'reward state' we can think more clearly, are more willing and able to collaborate, and are more open to new ideas and to change. Kawamichi *et al* (2015) used fMRI and found that being listened to activates the ventral striatum, part of the reward network.

## Insight

Chapter 4 touched on the importance of this but it bears repeating: organizations hugely underestimate the importance of allowing employees time to come to their own insights about why change is needed, what needs to change and how that will be done. Telling people to change leads to a threat warning in the limbic system and even well-meaning advice can lead to a threat response. Reaching insight is rewarding to the brain and avoids activating the limbic system. People need to be able to make their own connections and reach their own insights. When we reach our own insights we become more committed to an idea. The challenge in planning communication about change is in finding the right balance. If you give people too much detail, you prevent them from bringing their own insights and you reduce their level of autonomy. The challenge is in balancing this need for autonomy with our need for certainty. Most organizations tend to go into broadcast mode rather than allowing time for people to reflect, think and reach insights.

## Images

If you were to stop reading right now and look around you, you would be able to retain that picture far more easily than if I asked you to read a written description of it. Vision has evolved over millions of years and is a more efficient way of storing and remembering information than words. In short, we remember visuals better than words.

In his *Harvard Business Review* article (2015), Andrew Carton suggests that leaders who use image-based words in their vision statements are more successful at conveying the vision and in galvanizing people to work together towards a shared goal. Carton cites Churchill's 'we shall fight in the fields and in the streets', and Gates's 'a computer on every desk and in every home' because these provide images that people can imagine. Colleagues and he argue (2014) that clear images are particularly useful when talking about the future because the future is full of so much uncertainty. Images are more concrete than abstract concepts and values. Images give people something that they can imagine, envisage and work towards.

## Emotions

Intensely emotional experiences are more likely to be remembered because emotion increases attention. Emotionally-charged events register strongly in the sensory cortices and hippocampus. The advertisement for a charity mentioned in Chapter 6 is more memorable because it was about one child and therefore had a more emotional component to it than if the ad had been about facts and figures of damaged children. Most of us can remember where we were when we heard about the planes hitting the Twin Towers in New York in September 2001 because it was such an emotionally-charged event. The images of that day will probably be ones that we can recall for ever. Moments such as these are known as 'flashbulb memories': flashbulb memories are characterized by highly emotional events, and we perceive the memories as vivid and accurate, although they are often not so accurate in reality (Talarico and Rubin, 2003). Communication with an emotional element is more likely to catch our attention and to stay in our memories. Not that these emotions need to be unhappy ones for us to pay attention. The move to encourage leaders to tell personal stories is a reflection of the need not just to rely on facts and logic in organizational communication, but to share some emotion.

## *Involving people in change*

Organizations now have many means of communicating with, and involving people in, change. The introduction of social media into organizations means that thousands of employees can quickly provide their thoughts and ideas. Employees are a hugely valuable source of information for their employers. Many of them speak to customers and other stakeholders and can often see what needs to be put right. Involving people takes time and needs to be done with caution as it can raise expectations which then might be disappointed. Some types of decision clearly don't lend themselves to consultation with all staff – for example, which companies to acquire, which IT system to introduce. Leaders also need to be honest with themselves and with their people about where they are genuinely seeking input that might influence a decision and where they might be listening, but what employees say will have little impact. There's a temptation to tell employees they are being consulted when leaders already know what they want to do. Involving people – giving them some influence – can be very useful both in reaching the right decision and in building commitment to the changes.

## Autonomy

No one likes to be micro-managed. Having some control, or at least feeling we have some control, is important at work and for life in general. Judith Rodin and Ellen Langer (1977) examined the effect of choice on older people who lived in a nursing home. Residents were randomly assigned to two groups: one group was given control over their living environment, being able to choose their furniture, spend time with whom they wanted and were asked to look after a plant. The other group was not given this choice or responsibility. After 18 months, the group that had control had better health, physically and mentally, and were more likely to still be alive. Other studies have supported these findings and they have proved influential in changing how old people are cared for. For these elderly people, having some sense of control had a significant impact on their well-being. Elderly people are not the only ones deeply affected by a sense of control – we all are. Pain relief is an area where it is now common practice in many countries to give patients control. This is due to research that has shown that giving patients control can either lead to their taking less morphine (Mackie et al, 1991) or that their satisfaction with pain relief is better (McNichol et al, 2015).

A sense of control influences how we perceive events. We can look at the same situation and feel quite differently about it if we have some control. Even a subtle sense of control can significantly change our brains' perception of events. In an organization that was going to relocate, employees were very unhappy about the move and were vocal about their dissatisfaction. After a survey that simply asked people to choose what colour chairs they would like, complaints about the relocation noticeably diminished. Leaders and managers need to keep reflecting on where they can let go and give people choice. Choice reduces stress and cortisol. In Chapter 4 we looked at Yerkes and Dodson's inverted U of performance (see Figure 4.2). Amy Arnsten found in her research (2009) that one of the factors that can bring people back to the centre of the inverted U if they are stressed (and so have gone too far to the right of the U) is a sense of control, or at least a perception of choice. We have more resistance to stress when we feel we have some control.

Events that are uncontrollable and unpredictable are very stressful to us. It's part of the reason why terrorist attacks and domestic violence can be psychologically so damaging. At the much less serious end of the spectrum – unexpected traffic jams that mean we will be late for an important meeting, delayed trains with no communication as to when it will be on the move again, teenage children who won't answer their phone when they should have been home hours ago, a boss that is an unpredictable maverick – these are stressful to deal with. Just as a side note, giving choice to children between two options that are both acceptable to the parents can be particularly useful, eg 'Do you want to go to the cinema on Friday night or Saturday night?' Children have a less developed prefrontal cortex and so cannot handle lack of control as well as most adults. In fact, the PFC does not fully develop until our mid- or late twenties, which is a point to note for managers with people in their twenties in their teams.

## Storytelling

I was part of a team that worked on a takeover in the United States. Initially, this was not a particularly happy takeover as the acquired company was taken over by a company whose reputation had been badly damaged. We worked with the Sales Director and introduced him to the change curve. This is based on Elisabeth Kübler-Ross's research into the emotions we experience when faced with death (1969), and later established to be the emotions caused by any significant change. The change curve gave the director an insight into what he had been going through – the initial downward part of the curve where we experience denial, anger, and a desire to return to the past, and then with time, acceptance and an ability to think about, and plan for, the future. He decided to use it at a town hall meeting with his sales team. Using the curve, he described how he had felt about the takeover and what he had been experiencing. Referring to the lowest part of the curve, he said he felt it was time to make a decision and he had decided to stay. He then asked the team to use the curve to reflect on their emotions and to look to the future and to decide whether to stay in the new company or to leave. He wanted them to stay but understood if they felt they had to go. It proved a turning point for him and his team. Without being aware of the techniques, he was using both storytelling and emotional regulation (see Chapter 6) to help people deal with the changes. He used the change curve to 'label' his emotions which dampened down the amygdala. The very fact of observing our emotions helps to change them. He also used reappraisal of the situation, choosing to look at it in a different light and see the positives. Reappraisal is recognized as one of the best ways to manage our emotions

– see Gross and Ochsner's five stages of calming emotions in Chapter 6. He reminded people that they had a choice to go or stay: this gave them a sense of control, which in itself helps to reduce cortisol and stress. In addition to all of that, he was telling people a story about himself. As we will see, telling a story is a means of presenting an idea in a non-threatening way. This section will take a look at why storytelling can be both useful and persuasive during change.

## Why have we evolved to like stories?

Our brains are helpless storytellers. Heider and Simmel's simple animation (1944) of two triangles – one large and one small – and a small circle inside a 'box' demonstrate how we want to create stories. At first the large triangle seems to bully the smaller one, then it goes after the circle, and eventually the small triangle seems to rescue the circle. They are just shapes being moved around and yet our brains can't help turning them into a story, giving meaning and intentions to them.

From an early age, we like stories. Why have we evolved to like them? What advantage does it give us? As we know from earlier chapters, much of what drives our brain is a need to be able to get on with others. To do this we need to be able to understand others, recognize that they have different thoughts from us and try to comprehend what they might be feeling, thinking, and intending to do. The ability to do this is known as 'theory of mind' (ToM) or 'mentalizing'. Reading or listening to stories taps into ToM, and allows us to step into the shoes of others and explore the world from their perspective. In understanding fictional characters, we use the same processes as for understanding real people. Stories improve our ability to understand others and to get on with them. Various studies, including research conducted by Mar *et al* (2010), support this idea: in preschool children there is a link between exposure to storybooks and social development.

One advantage of storytelling from an evolutionary point of view might be that they enable us to explore simulations and different scenarios and get better at predicting outcomes. This helps us to prepare for changes that we have not yet personally experienced. The benefits of this during organizational change are obvious.

## Why are stories persuasive?

When we make a presentation, we tend to fill our slides with bullet points and facts and figures. These activate certain parts of the brain involved in processing language, in particular Broca's area and Wernicke's area (see Chapter 1) and they activate the analytical parts of our brain. As we listen to the presentation, we are likely to critique what we hear and question whether we believe it, agree with it, etc. Storytelling does something different to the brain. It activates not just language networks but also ToM networks, including areas such as the superior temporal gyrus/sulcus and the temporoparietal junction. In some cases, it also activates the inferior frontal gyrus which is known to be a 'mirror neuron' system. As we saw in Chapter 6, this means that the same neurons are activated as if we were executing an action ourselves. Research has been conducted by, amongst others, Véronique Boulenger (Boulenger *et al*, 2006) to show that if we hear about someone kicking a ball, for example, our motor cortex is activated. When we hear about coffee and perfume, the olfactory cortex is activated; hearing about emotions activates our emotional network; even reading about dialogue activates the auditory part of our brain. Hearing a story activates far more parts of the brain: it's as if we are in the shoes of the people in the story and are experiencing the sensations and emotions that they are experiencing. Told in the right way, a story is a much less threatening way of planting an idea in someone's head than telling someone that they need to change. We experience and identify with a story, whereas we filter and analyse a presentation.

Evidence (Green and Brock, 2000) suggests that after reading or listening to stories, our beliefs and attitudes shift in line with the narrative. The more engaged people feel in the story and the more vividly they experience it, the more they adapt their beliefs. In particular, if we empathize with the character and that character changes their beliefs, we are more likely to change our minds. Thinking again about the popular Sales Director, that's exactly what he did with his salesforce.

A few last points about stories: they are memorable. If you ask people brought up in a Christian environment if they can tell you, for example, the Nativity story and the Ten Commandments, the chances are that they will be able to tell you some of the former but not many of the latter. Narratives are easier for us to remember than lists of points because they have a beginning, middle and an end. There are causes and effects, ie a pattern to them, and our brains like patterns. Often there is a novelty element to stories, and our brains like novelty: it catches our attention. Finally, stories allow us to reach our own insights and, as we know, insights are rewarding to the brain and help build commitment to an idea.

# What can we do? Solutions and examples of what other leaders have done

Communication during change is vital: change won't happen without it. Change is difficult for us because it requires more energy than to stay as we are. Change itself creates a threat state and cognitive dissonance because it suggests our old practices were wrong. We need to talk about the past with respect and plan and deliver communication with the threat and reward states in mind. Here are some practical examples of what organizations have done.

## Provide certainty about communication

As we know, our brains want certainty and this can be difficult to provide during change. To help improve certainty, some communication teams commit to how they will communicate with employees during change. For example, this could be guaranteeing that there will be weekly updates even if the update is to say there is no news. They develop communication principles setting out their promise so that, in an uncertain world, employees can at least be sure of the communication process.

## Enable insight

We are more inclined to support an idea when we have had a moment of insight. Rather than telling people why change is good, some organizations now focus on conversation. This means equipping leaders with the skills to have a dialogue, to ask questions and not to feel uncomfortable when they don't have all the answers. One large bank has completely changed its team meetings such that managers now just listen. This means that they are seeing all sorts of ideas being suggested by employees, and employees report feeling more positive.

'Big' or 'rich' pictures or 'learning maps' are a good means to get conversations going. These are large pictures which can be put up on the wall and employees gather round. Managers are trained to use them not as a presentation tool but as a means to spark conversation about where the organization is heading, what challenges lie along the way, what benefits there will be, what role each team and each person needs to play. Big pictures, as the name suggests, use many images rather than words and so they are visually appealing. The fact that employees gather round them with the facilitator means that the focus is on a discussion and sharing of ideas. Once employees have held the initial conversation, teams can come back to the picture at a later stage to discuss progress and the pictures can be displayed around the workplace as a reminder of the conversation.

## Involving people in change

Involving people in change can improve the quality of decision-making and helps to build commitment. There are many ways in which this can be done both on a small, local scale and right across the organization. Social media have particularly helped with the latter. Following a merger, one US organization consulted several thousand leaders globally online on what the values of the company should be. Through facilitated conversations, they allowed the executives to air their concerns and irritations before involving them in a more constructive discussion. The result was that the values had been developed with the input of many leaders and these people felt ownership of the values before they were even launched.

When there is large-scale downsizing, managers often find it difficult to involve their staff: jobs are going, sites are often closing, there are not many options – what is there to involve people in? In a large UK government department, managers were aware of the need to give some control to employees. Some focused on helping employees to take control of their future by running workshops on the transferable skills employees had acquired over the years. They developed a personal 'moving on planner' that got employees to think about what they wanted, for example:

- I'm packing...
- I'm dumping...
- What I like about where I am now...
- What I don't like about where I am now...
- What is going to be hard...
- How I plan to deal with those challenges...
- What impression I want to make with my new colleagues...
- What I'm good at...
- I plan to make it a good experience by...
- What I will do differently with my new team...

- What skills do I want to use in the future?
- What will I be glad to leave behind?

This gave people a means to plan and a sense of some control over the future.

Another useful exercise is to help people identify what they can't control and to focus on what they can control. This can either be done in columns or within circles. Draw a large circle, then another one within it, and a smaller one inside that one. In the inner circle, people write what they can control; in the middle circle they note what they can influence or partially control with the help of others; and in the outer circle, they list issues that they cannot influence. They then agree where they can most constructively focus their efforts. As we saw earlier, involvement in even seemingly small-scale decisions, such as what colour chairs people want, can make a difference to people's sense of control and therefore to how they see the changes.

## Explain why storytelling works

Many organizations are asking leaders and managers to use more stories in their communication. Some leaders and managers don't know why they are being asked to do this. A UK-based financial services organization ran sessions for their managers on the neuroscience behind storytelling so that they knew why they were being asked to do this. One useful exercise was to ask managers to prepare a piece of communication, firstly by using a page of bullet points, and then secondly by providing an anecdote or story that brought the points to life. Both the managers and their listeners were asked to feedback how different the two experiences felt. This enabled them to experience the difference in approaches.

## Using visuals

There are lots of examples of infographics and 'rich pictures', 'big pictures', or 'learning maps'. Big pictures have the advantage of not only being visual and therefore attention-grabbing and memorable: they also encourage people to talk to one another. Many organizations have used them with great success as they bring the vision to life. In fact, the danger of them can be that they work so well, that every part of the organization wants to create one, and suddenly there are pictures everywhere. They are a great means, not only to convey images of what the future looks like, but also to tell stories, share ideas and plan what steps to take. They can be complete or semi-complete and teams fill in the gaps. They work well in multilingual companies as there tends to be just one picture, ideally with very few or no words. This picture unifies all employees and is equally accessible to all because language is not a barrier. Reminding people of shared goals is a great means of focusing on what people have in common across the business and is a means of creating an 'ingroup' as we saw in Chapter 5.

## Threat or reward: what impact are you having?

Some communication teams run a 'neuroscience audit' of their communication practices and messages, identifying what impact they might inadvertently have. They want to understand where they might be causing a threat state and what they can do to help keep employees motivated and constructive. Very specifically, one leader in financial services, having understood about the stronger impact of the threat state than the reward or toward state, asked her managers to come to a meeting with the last 10 e-mails they had sent. They critiqued them against threat and reward and identified what impact they thought each e-mail would have. They also thought about timing and considered the consequences of sending an e-mail that would cause a threat state to a manager, eg just before that manager was about to run a team meeting. If the manager has just been tipped into a threat state, what impact would that have on the team? Leaders need to deliver difficult messages but they need to think about the consequences and the best timing.

## Be visible

One of the most successful CEOs I have come across was hailed by *The Financial Times* (*FT*) as 'the turnaround king' and he increased the company's share price eightfold. He declared that any day in head office was a day wasted. He understood that change requires leaders to connect with people and spent as much time as he could visiting sites scattered around the globe. He spent his time in conversation with employees. He said he had seen change efforts fail in previous organizations because leaders did not spend enough time talking with employees. He understood our need for social connection (see Chapter 5). As we know from previous chapters, our brains like familiarity and we tend to trust familiar faces more than unknown ones – being visible makes a difference. As the *FT* recognized, and the share price confirmed, this CEO's efforts paid off.

## Cognitive dissonance is distracting

As we know from personal experience, people hear different things. Neuroscience explains why this happens – different context, emotions, experience, expectations and biases filter and distort what we take in and how we feel about what we have heard. Uncertainty means that people's brains are distracted and are less capable of absorbing information. Cognitive dissonance means that people will not be listening properly, so if the leader is announcing something surprising or that does not seem to fit with what has been said in the past, employees will be distracted until they can make sense of it. Repeating messages, putting them in writing and checking for understanding are key.

## Transparency

Our needs for transparency and fairness are always important to us, but this is all the more true when we are going through periods of change. There are many ways of achieving this. One manager in a UK government department put a large piece of paper on the wall and asked the team to write any questions on it and then he tried to get answers to the questions each week. Others have talked about having 'myth-busting' sessions, where they addressed rumours and speculation and clarified what was happening and what was not on the agenda. Other organizations have got their CEOs to use social media to address questions personally.

## Emotions

Organizational communication tends to rely on logic. Having well-reasoned arguments and evidence as to why the organization and employees need to change is important. But the brain tends to pay attention and remember messages that stir emotions. Emotions help us to retain information. The Sales Director using the change curve is in part memorable because he told a very personal story. He didn't provide a list of rational reasons for leaving or staying with the company. He spoke from the heart.

## Purpose

Under stress, we tend to remember negative messages, so it is important during change to focus on positive future possibilities. We saw in Chapter 5 the importance of developing the ability to reappraise and reinterpret to reduce the amygdala's response. People benefit from doing this on a personal level but internal communication needs to help people do this. In Chapter 5, we looked at the research conducted by Adam Grant (2014) amongst fund-raising teams and the remarkable difference it made when employees are put in touch with the beneficiaries of their work. They had been reminded of why they came to work and the difference that they made. They knew their purpose.

This is important in change: employees need to be clear about what the purpose of the change is and what benefits it will bring. Big pictures are very useful in that, once they have been used by teams, they can be displayed around the building or electronically, acting as a constant reminder of the organization's purpose – why do we exist, what difference do we make?

## What's changing and what's not

When going through change, there is a tendency to focus on what will be different. But given the need our brains have for certainty and the fact that they like familiarity, it is also important to remind people of what will stay the same. If the values still hold true, the commitment to customer service or to innovation remains valid – employees need to be reminded of this so that they recognize not everything is in a state of flux.

## A few other points in brief

When planning communication during change, there are a few other points from previous chapters to keep in mind:

- When people are struggling with change, using communication to remind them of when they have done things well in the past can help them feel more positive (see Chapter 4).
- Recognition is important and unexpected recognition generates even more dopamine (see Chapter 4).
- Our brains like a bit of novelty (see Chapter 4).
- Laughter, used appropriately, can help people get through difficult times (see Chapter 4).
- Our brains are limited capacity machines and there is a limit to how much we can take in, so simplifying complicated ideas helps people's brains (see Chapter 7).

- Our brains tire easily and glucose levels make a big difference to our ability to pay attention, so think about timing of the day. If these are complex or sensitive messages to convey, communicate these earlier in the working day (see Chapter 7).

# Summary of key points from this chapter

## *The science*

- Our brains are constantly in search of information and the sooner we get it the better.
- Even bad news can be preferable to no news as it gives certainty and allows people to plan.
- Lack of information causes an 'error alert' in our brains.
- This hunger for certainty can mean that we sometimes grasp at a solution too quickly when it would be better to live with uncertainty for longer.
- Communicating sooner or later in a change programme: both present challenges. If we communicate early, we might need to correct information. This causes problems because we put pressure on our leaders not to change their minds or to be wrong. If leaders wait until there is more certainty to communicate, there will be a communication gap and employees will waste energy speculating.
- What each of us sees as reality is highly subjective.
- Past experience, current context, personality, expectations and biases influence what we perceive and what we take in.
- We need to set goals carefully: not meeting our expectations leads to a drop in dopamine and increased activity in the amygdala.
- Meeting and exceeding expectations leads to an increase in dopamine.
- Our brains do not like to experience conflicting ideas (cognitive dissonance) – it activates the pain network, creates an 'error alert' and means that our brains are distracted until we have resolved the conflict.
- The brain will try to resolve the conflict of ideas by removing one of the beliefs or by creating a new one.
- We like to live up to the image both we and others have of ourselves (ie avoid cognitive dissonance). This is one of the reasons why public statements can increase commitment – we feel we need to live up to what we have said.

- We don't take in as much as we think we do: our brains have blind spots and a limited capacity. If we focus too much on one area, we miss the bigger picture.

- Uncertainty leads to a threat response in our brains which means that our sensitivity to perceived threats is heightened. Even well-intentioned words and actions can trigger a threat response, especially when we are feeling uncertain.

- There are conflicting views as to the impact of screen time: more evidence is needed.

- There are two key traits that we notice in leaders: warmth and competence. Research suggests leaders should focus on warmth first as this leads to trust and leaders being listened to during change.

- Positive social interactions, including being listened to, activate the reward network and put our brains into a more positive place.

- Organizations need to enable employees to come to their own insights about the need to change: insight is rewarding to the brain.

- Visual communication is an efficient way of taking in and remembering information.

- Emotions capture our attention.

- No one likes to be micro-managed.

- Events that are uncontrollable and unpredictable are very stressful.

- Autonomy is important for us mentally and physically: a perception of control reduces cortisol and stress.

- Even small choices make a big difference to us.

- Our brains like stories and they enable us to develop our social skills and test out new scenarios and how best to respond to them.

- Hearing a story activates many parts of our brain: we experience the story with the storyteller.

- Stories are more memorable than a list of points.

- Stories can be very persuasive.

## *What can we do?*

- Be careful how we position change: change suggests that what we are currently doing is wrong.

- Don't be surprised by people hearing different things and responding differently: our perceptions are highly subjective.

- In an uncertain environment, provide certainty about the communications process.

- Enable insight: ask questions, create dialogue.

- Help people to focus on what they can control; involve them in decision-making where appropriate.

- Explain to leaders why storytelling works before asking them to use the technique.

- Use visuals.

- Equip leaders to use big pictures to hold conversations and to share superordinate goals.

- Threat or reward: think about what impact your communication will have – conduct a communications audit testing your communications against threat/away and reward/toward responses.

- Remember that people in a threat state cannot take in so much information – their brains are distracted and more sensitive to negative messages.

- Be visible: we have a strong need for social connection and we tend to like familiar faces.

- Plan for cognitive dissonance – repeat messages that might have been missed.

- Provide transparency – enable people to ask questions and get answers.

- We tend to use logic in organizational communication but emotions are more memorable.

- Remind people of their purpose and what difference they make to others.

- Be clear about what is changing and what's not.


## Reflections and planning

We know that the brain craves certainty, so:

- What more can you tell people?

- Do you have regular communication times and processes in place to help the brain predict?

- Are there times when you might be jumping to solutions too quickly?

- Do you explain the rationale behind important decisions that affect people, wherever possible?

We know that the threat state is easily triggered in the brain, so:

- Have you reviewed your communications to see what state of mind they will create?
- Have you thought about the timing of sending difficult or negative messages and their impact on people's ability to focus?
- Are leaders aware of how they might unwittingly cause a threat state through words or actions?

We know what each of us perceives is highly subjective and this affects what we hear and how we respond to change, so:

- Managers are key in helping individuals deal with change: are they aware of why we see and hear different things?
- How can you help managers understand what might be influencing their perceptions of change?
- How can you best help managers to deal with the range of reactions to change within their teams?
- What rumours and misinformation lurk in the organization? Do you know? How can you address them?

We know that a sense of some control has a major impact on reducing stress, so:

- What are you already doing to involve people? Can you do more?
- Where else can you pass down responsibility?
- Are managers aware of the importance of choice? How can you help them let go of control?
- What happens when leaders are under pressure – is there a tendency to micro-manage? How can you help leaders better manage emotions and continue to delegate?

We know that storytelling can be persuasive, so:

- What stories will bring more meaning and clarity to the changes?
- Do your leaders and managers understand why storytelling works?
- Are they confident enough to tell stories to bring communication to life? If not, what support do they need?
- What opportunities are there for all employees to tell their stories about the impact of the changes?

We know that visuals are a highly efficient way of communicating, so:

- Where are the opportunities for you to use visuals in your change communication?
- How can these images be used again to act as reminders of key messages?
- What language do your leaders use? Do they use images when they speak or are they abstract?

We know that insight is rewarding to the brain and helps build commitment to change, so:

- What opportunities are there for dialogue in your organization? How can you create more opportunities?
- Are you getting the balance right between providing information and allowing employees to make their own connections?
- How can you persuade leaders of the need not just to be in 'broadcast mode' but to allow people time to reach insight?

We know that having a sense of purpose and doing good for others is hugely motivating to our brains, so:

- When did you last communicate the direction and purpose of the organization?
- When did employees last meet people who benefit from their work?
- Does everyone recognize the value of the contribution they make and feel appreciated?
- Do you reward the activities and behaviours that really matter?

# References and further reading

Arnsten, A F T (2009) Stress signalling pathways that impair prefrontal cortex structure and function, *Nature Reviews Neuroscience*, **10** (6), pp 410–22

Baumeister, R F and Leary, M R (1995) The need to belong: desire for interpersonal attachments as a fundamental human motivation, *Psychological Bulletin*, **117** (3), pp 497–529

Boulenger, V *et al* (2006) Cross-talk between language processes and overt motor behavior in the first 200 msec of processing, *Journal of Cognitive Neuroscience*, **18** (10), pp 1607–15

Bromberg-Martin, E S and Hikosaka, O (2009) Midbrain dopamine neurons signal preference for advance information about upcoming rewards, *Neuron*, **63** (1), pp 119–26

Carton, A M (2015) People remember what you say when you paint a picture, *Harvard Business Review*

Carton, A M, Murphy, C and Clark, J R (2014) A (blurry) vision of the future: how leader rhetoric about ultimate goals influences performance, *Academy of Management Journal*, 57 (6), pp 1544–70

Cialdini, R B (2001) *Influence: Science and Practice*, Allyn and Bacon, Needham Heights MA

Cioffi, D and Garner, R (1996) On doing the decision: effects of active versus passive choice on commitment and self-perception, *Personality and Social Psychology Bulletin*, 22 (2), pp 133–47

Cuddy, A J C, Kohut, M and Neffinger, J (2013) Connect, then lead, *Harvard Business Review*, July–August

Festinger, L, Riecken, H W and Schacter, S (1956) *When Prophecy Fails: A social and psychological study of a modern group that predicted the destruction of the world*, University of Minnesota Press

Galinsky, A D *et al* (2006) Power and perspectives not taken, *Psychological Science*, 17 (12), pp 1068–74

Grant, A (2014) *Give and Take*, Orion Publishing, London

Green, M C and Brock, T C (2000) The role of transportation in the persuasiveness of public narratives, *Journal of Personality and Social Psychology*, 79 (5), pp 701–21

Heider, F and Simmel, M (1944) [accessed 8 August 2015] Heider and Simmel animation [Online] https://www.youtube.com/watch?v=VTNmLt7QX8E

Kawamichi, H *et al* (2015) Perceiving active listening activates the reward system and improves the impression of relevant experiences, *Social Neuroscience*, 10 (1), pp 16–26

Koyama, T *et al* (2005) The subjective experience of pain: where expectations become reality, *Proceedings of the National Academy of Sciences of the United States of America (PNAS)*, 102 (36), pp 12950–55

Kübler-Ross, E (1969) *On Death and Dying*, Routledge, London

Lieberman, M D and Eisenberger, N I (2009) Neuroscience: pains and pleasures of social life, *Science*, 323 (5916), pp 890–91

Mackie, A M, Coda, B C and Hill, H F (1991) Adolescents use patient-controlled analgesia effectively for relief from prolonged oropharyngeal mucositis pain, *Pain*, 46 (3), pp 265–69

Mar, R A, Tackett, J L and Moore, C (2010) Exposure to media and theory-of-mind development in preschoolers, *Cognitive Development*, 25 (1), pp 69–78

McNicol, E D, Ferguson, M C and Hudcova, J (2015) Patient-controlled opioid analgesia versus non-patient controlled opioid analgesia for postoperative pain, *Cochrane Database of Systematic Reviews*, Issue 6

Rodin, J and Langer, E J (1977) Long-term effects of a control-relevant intervention with the institutionalized aged, *Journal of Personality and Social Psychology*, 35 (12), pp 897–902

Simons, D and Chabris, C [accessed 8 August 2015] Selective Attention Test [Online]
    https://www.youtube.com/watch?v=vJG698U2Mvo
Talarico, J M and Rubin, D C (2003) Confidence, not consistency, characterizes
    flashbulb memories, *Psychological Science*, **14** (5), pp 455–61
van Veen, V *et al* (2009) Neural activity predicts attitude change in cognitive
    dissonance, *Nature Neuroscience*, **12** (11), pp 1469–74
Wiggins, S *et al* (1992) The psychological consequences of predictive testing for
    Huntington's disease, *New England Journal of Medicine*, **327** (20), pp 1401–05
Wiseman, R [accessed 8 August 2015] (uploaded 2012) The Amazing Colour
    Changing Card Trick [Online] https://www.youtube.com/watch?v=v3iPrBrGSJM
Yerkes, R M and Dodson, J D (1908) The relation of strength of stimulus to rapidity
    of habit-formation, *Journal of Comparative Neurology and Psychology*, **18** (5),
    pp 459–82

# Planning change with the brain in mind

## 09

Change is hard to plan because, by its very nature, we don't know what will happen during the course of the changes. Organizations need to have direction but also be flexible. More important than a change plan, is to have a change planning process that allows the organization to track progress, take stock and remain agile.

## Change requires more energy than staying as we are

Change is difficult for the brain as it prefers to be on automatic pilot and to continue with its old habits. Change causes activation in the amygdala and the limbic system and is associated with error detection and threat. In fact, research conducted by Herry *et al* in 2007 demonstrates that unpredictability in itself is enough to unsettle us. Both humans and mice, when exposed to unpredictable noises that had no significance, were shown to have more activation in the amygdala and behave in a more anxious way. This just goes to show the depth of our intolerance for uncertainty. Our brains want to conserve energy and change requires more energy than staying as we are. The challenge for organizations going through change is that it creates a threat state. Figure 4.1 in Chapter 4 sets out some of the consequences of the threat state. The diagram is included again (see Figure 9.1) as it is an important reminder as to why organizations should be concerned about what state their employees' brains are in. On the left is the brain of an employee that really can't perform well. On the right is the brain of an employee who is open to new ideas and more capable of dealing with change. There are many different descriptions of 'employee engagement' but the mindset described on the right-hand side is clearly that of someone who is engaged at work.

**Figure 9.1**    The impact of threat and reward on our brains and on our ability to think and perform

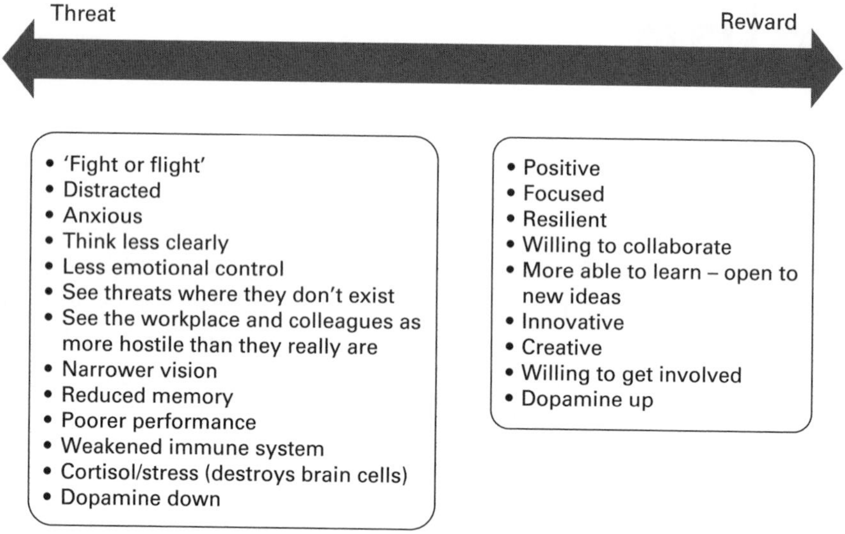

Threat                                                                                Reward

| |
|---|
| • 'Fight or flight' |
| • Distracted |
| • Anxious |
| • Think less clearly |
| • Less emotional control |
| • See threats where they don't exist |
| • See the workplace and colleagues as more hostile than they really are |
| • Narrower vision |
| • Reduced memory |
| • Poorer performance |
| • Weakened immune system |
| • Cortisol/stress (destroys brain cells) |
| • Dopamine down |

| |
|---|
| • Positive |
| • Focused |
| • Resilient |
| • Willing to collaborate |
| • More able to learn – open to new ideas |
| • Innovative |
| • Creative |
| • Willing to get involved |
| • Dopamine up |

As we know from earlier chapters, the threat state might be strong and employees might feel anxious and be less able to function well at work. It can also work at an unconscious level when the amygdala is activated and we are not aware of it. This is particularly true when it is not a major change programme, but constant small levels of change are required. Significant change programmes catch our attention and organizations plan for them. But for most people, there is constant change in the workplace – new systems to learn, new legislation to deal with, a new CEO, a new team, cost-cutting initiatives and so on. Our brains like a bit of novelty and change – boredom is fatiguing – but lots of small changes are likely to elicit a threat response. As we saw in Chapter 8, change that is unpredictable and uncontrollable is very stressful. We need to help people to be able to see what is on the horizon and to feel they have some control over what is happening to them. Planning change with the brain in mind is all about staying aware of what will cause a threat state. Leaders need to think about both the macro activities and the micro activities – from the major decisions about change through to acknowledging a team member as they pass them in the corridor. It is also about identifying the large or small actions that will engender a reward state, where employees' brains are able to focus, learn and collaborate.

# Change models

There are many models of change that guide how leaders should think about change and plan for it. In this section, we will take a look at a few of them and then explore the neuroscience behind them.

## The four enablers of engagement

This model was not specifically designed for change but does focus on what helps employees to perform at their best. Based on extensive research conducted by David MacLeod and Nita Clarke for the UK's *Engage for Success* movement, it is known as the MacLeod Report (2009). The report set out to identify whether 'employee engagement' makes a difference to organizational performance and, if so, what conditions need to be present to lead to employees' feeling engaged. The answer to the first point was an unequivocal 'yes': there are strong links between the two. They found that in organizations where employees felt positive about the workplace there was lower absenteeism, improved customer service, and advocacy.

The report identified four factors that distinguished those organizations with high employee engagement. They are:

1 Visible, empowering leadership providing a strong strategic narrative about the organization, where it's come from and where it's going.

2 Engaging managers who focus their people and give them scope, treat them as individuals, and coach and stretch them.

3 There is employee voice throughout the organization, for reinforcing and challenging views, between functions and externally; employees are seen as central to the solution.

4 There is organizational integrity – the values on the wall are reflected in day-to-day behaviours. There is no 'say–do' gap.

These four 'enablers' make sense from a neuroscience perspective. The first states that organizations need to have a strong strategic narrative. Having a clear sense of direction provides both a sense of purpose and certainty. Having a clear story about where the organization is headed enables employees' brains to predict, and we know from Chapter 2 how important prediction is to the brain. Engaging managers, the second 'enabler' touches on many of the points we have covered in other chapters. We know from Chapter 5 on the social brain that we work much better when we have managers who accept us and treat us with warmth and empathy. Managers who focus people and give them scope are the ones who find the right balance between certainty and autonomy. We saw how important these were in Chapter 8 – the two are closely linked and both help to reduce distraction and the threat state. A manager who coaches is a manager who asks questions, and we saw in Chapter 8 that this activates the reward circuit. Asking questions should also mean that the manager is interested in the view of the employee: believing that someone has a good opinion of us is rewarding to the brain. A manager who coaches should also be a manager who believes in Carol Dweck's mindset – that ability is not fixed at birth, that we can get better at things if we apply ourselves and are prepared to learn from our mistakes. A coach is also someone who lets the employee find the answer, nurturing autonomy. Stretching people means that they will help them find the right place on the Yerkes–Dodson inverted U of performance (1908) – they are challenged but don't feel out of their depth. From what we know about the reward system in the brain, this enabler 'ticks a lot of the boxes'. The third, employee voice, is all about enabling people to be heard and to feel they have some influence. We saw the importance of autonomy in Chapter 8. The fourth, organizational integrity, means that what the organization says it values really is valued, recognized and rewarded. Espoused and real values are the same. In an organization where this is the case, employees will have a greater sense of certainty. They know what they need to do and how they should behave to get recognition. This fourth enabler also promotes transparency and fairness, which activate the reward network.

Neuroscience helps to explain why these four enablers of engagement create an engaged mindset. An organization where employees felt unsure of their direction and how their contribution made a difference, where managers were distant, uninterested and disrespectful, and where employees felt they had no influence and could not trust what the organization said it valued, would be an organization where employees would be in a very strong threat state and their brains would be unable to think or perform well.

## Autonomy, mastery and purpose

In his book, *Drive: The Surprising Truth About What Motivates Us* (2009), Dan Pink sets out three factors that motivate people – autonomy, mastery and purpose. These are intrinsic motivators, ie they are not driven by carrots or sticks that the company might choose to provide. As Pink recognizes, carrots and sticks tend to create compliance, and when the carrot or stick is removed the desired behaviour stops. The desire for autonomy, mastery and purpose comes from what we find internally rewarding. Autonomy and purpose are covered in the Four Enablers in this chapter. The third component, mastery, is about the desire to improve. Pink describes Csikszentmihalyi's concept of flow (1990) that we looked at in Chapter 4 – those moments when we are completely absorbed in our work, when the activity itself is so enjoyable we don't need any other reward. The inverted U of performance is relevant here too: when we are in flow we are at the pinnacle of that inverted U – the activity stretches us and requires effort but it is not so difficult that we are anxious. Feeling that we are improving at something activates the reward circuitry. Mastery also touches on Carol Dweck's research (2007) on fixed and growth mindsets (see Chapter 4). With mastery the enjoyment is in learning and improving, not trying to demonstrate how capable we already are. Pink says that mastery is not all pleasure though: it is also about having grit, perseverance and determination to reach a goal. In part, this is down to personality but, as we know from Baumeister's, and Eisenberger's and Lieberman's work (Baumeister and Leary, 1995; Eisenberger *et al*, 2003), people are more likely to persevere if they are in the presence of an empathetic person, feel accepted and have control.

## Kotter's eight steps to transforming your organization

Kotter's well-known 1995 *Harvard Business Review* article, 'Leading change: why transformation efforts fail', sets out eight steps that organizations need to go through:

1 Establishing a sense of urgency.
2 Forming a powerful guiding coalition.
3 Creating a vision.
4 Communicating the vision.
5 Empowering others to act on the vision.
6 Planning for and creating short-term wins.

**7** Consolidating improvements and producing still more change.

**8** Institutionalizing new approaches.

The eight steps are self-explanatory, and from what we have explored from the fields of neuroscience and behavioural science in the earlier chapters, these steps make sense. We need a reason to change, a clear vision of where we are heading, and some influence in implementing the changes. As we saw in Chapter 4, short-term wins are important as achieving a goal generates dopamine and puts people's brains in a better place to take on the next challenge. Building on this, neuroscience would suggest that people need to have time to reach their own insights and to understand why the change is needed. It is better to enable people to make their own connections than to tell them the organization needs to change and so do the employees. The latter leads to a threat response and compliance; the former is rewarding.

## *Bridges' Transition Model*

William Bridges' model of change (1991) identifies phases people go through when dealing with change. Bridges focused on the psychological transitions that change causes. The model is not only useful in recognizing what is happening to people mentally and emotionally, but also as a practical guide as to what to do to help people through the confusion of change. He identified three phases: Endings, Neutral Zone and New Beginnings.

Endings is the first stage – when we realize that the old way of doing things is coming to a close. The model recognizes that people have to let go of the past before they can look forward to the future – a step many teams miss in their haste to set up new structures and get on with making change happen. Bridges' model is also useful in that he acknowledges the 'Neutral Zone' – a period where we know things are changing but we do not yet have much clarity about what the future looks like. It's that messy stage of change where things are still in limbo. As we know, the brain does not cope well with uncertainty or ambiguity, so Bridges suggests that people try to reframe the experience (see Chapter 6 on our emotional regulation) by looking at it as an opportunity to be creative. The old structures, rules and ways of working might be disappearing so, freed of these constraints, this is a time to think about new ways of doing things. Bridges finds the positive in what is so often seen as a negative time in organizations. Bridges also recognizes people need to be given certainty where possible: they need short-term goals that they can achieve and they need temporary structures and roles that provide a framework. He also talks about the need to protect people from failure. People's brains will be vulnerable to being in a threat state in the Neutral Zone, and any sense of failure would compound this and send them into a deeper threat state. So, the question here for leaders and managers is: how can they avoid the threat state? As we know, small things can tip our brains into that state. Leaders and managers need to think each day about what they can do to move people towards a reward state.

In 'New Beginnings', the last of the three phases of transition, Bridges suggests depicting the future. He too emphasizes the importance of visuals as they help to make the future more real, less abstract. As we saw in Chapter 8, our brains like visuals and they are a very efficient way for us to absorb information and to get clarity. We have already covered the importance of certainty to the brain and Bridges also stresses the importance of involving people. He emphasizes the importance of encouraging people to do things in new ways and in celebrating achievements. Working in different ways is hard for us initially, and it takes energy and practice for these new ways of working to become habits whereby we don't have to think so consciously about them. It is easy for us at this stage to slip back into old ways of doing things because they are more established in our brains and take less energy. As Bridges suggests, we need to encourage and reinforce new ways of working. There is much from what we understand about our brains that would support Bridges' Transitions model.

## SPACES – Hilary Scarlett

Some people find it useful to have an acronym to remember the factors that have a strong impact on motivation and employee engagement. One I use to sum them up is SPACES. It stands for Self-esteem, Purpose, Autonomy, Certainty, Equity and Social Connection. Take a look at Figure 9.1 again: when we feel we have these SPACES 'factors' our brains are in a reward state, but when we lack them our brains go into a threat state, with all the consequences for our ability to perform set out in the list on the left of the figure. From previous chapters, it will be clear why these factors influence us and how we perceive the workplace. The powerful impact of Purpose and meaning is well illustrated in Adam Grant's (2014) research that we looked at in Chapter 5. Pink identified Purpose's role in motivation in his book, *Drive* (2009). We covered the brain's need for Certainty and Autonomy in Chapter 8. Our brains want to predict and having Certainty means they can do this. Feeling that we have some control greatly reduces cortisol and stress. Social Connection refers to our brains' need to feel that we are accepted by others. The work of Eisenberger, Lieberman and Williams (2003), and of

Roy Baumeister (Baumeister and Leary, 1995) amongst others, discussed in Chapter 5, demonstrates how much of what we do is driven by our need to feel we belong.

We haven't yet looked at why 'Self-esteem' matters to the brain. I have used the term Self-esteem but it has a number of facets to it. Self-esteem is in part about our hierarchical status and feeling better than others. Professor Sir Michael Marmot's article (2004b) based on his book *Status Syndrome* (2004a) makes the link between status and length of life. Put simply, the lower our place in the hierarchy, then the shorter our life expectancy. His two Whitehall studies (begun in 1967 and 1985) followed UK civil servants over periods of time. They found that the more senior the UK civil servants were in terms of their job grade, the longer they were likely to live, even when controlling for income, education, smoking and lack of physical activity. There are various theories as to why this should be: lack of control and lack of ability to predict, for those who are further down the organization, are amongst them. Julianne Moore, Best Actress Oscar winner in 2015, joked that she was delighted to win, not least because Oscar winners have been shown to live about four years longer than nominees (Redelmeier and Singh, 2001b). Professor Ian Robertson of Trinity College Dublin (2012) hypothesizes that this might be because in some ways winning removes a certain pressure: even if one of your films bombs at the box office or with the critics, winners will always have the Oscar 'badge' to show that they are good at what they do. Intriguingly, and in contrast to both actors and Marmot's studies, amongst others, that show a link between achievement and longevity, screenwriters who win an Oscar have a shorter life expectancy than nominees (Redelmeier and Singh, 2001a).

Oscar winners and UK civil servants are not the only subjects of research into status and longevity. In Robert Sapolsky's article, 'The influence of social hierarchy on primate health' (2005), he notes that some dominant animals that have to keep fighting to maintain their power, African wild dogs for example, have high levels of physiological stress. But where domination is maintained through intimidation, where even looking the alpha animal in the eye could be problematic (eg rhesus monkeys), it is the lower ranking animals that suffer more stress. There are strong associations between hierarchy and poor health, Sapolsky concludes, with humans or animals who are in the 'wrong' rank, suffering poor health due to physical or psychological stress.

There is not much we can do about the hierarchy in organizations; not everyone can be the CEO. However, there are some aspects of Self-esteem that we can influence. People belong to multiple hierarchies, and employees who might be quite low in the organizational hierarchy might rank highly in others, eg a great fund-raiser for the organization's chosen charity or a coach for a children's football team outside the workplace, and we can choose to recognize this at work.

'Self-esteem' is not just about being more senior in a hierarchy – it is also about being better than others. So, enabling your team to feel better than teams in, for example, the competitor's organization would boost a sense of Self-esteem. Being associated with a winning team (actively or vicariously) not only has an emotional and cognitive impact but also affects us physiologically in that it changes our hormone levels. Bernhardt *et al* (1998) found that testosterone increased in fans of winning teams and decreased in fans of losing teams. Anyone who is, or is close to, a sports fan will have experienced the effect of winning and losing performances, the highs and the lows.

Focusing the work team on being better than others might not always be the most constructive approach. Intra-organization competitiveness will lead to reinforcing silos and outgroups within the organization and this will hinder collaboration. A better approach to boosting Self-esteem is to enable people to feel 'better than themselves' – in other words, to help people improve and have a sense of personal achievement. As we saw in Chapter 4, reminding people how much they have progressed at a task activates dopamine and the reward system. Learning and personal development are good ways to boost people's Self-esteem. This is useful to keep in mind, as there are often opportunities for employees to have new experiences and learn new skills. Encouraging them to do this during change might be particularly constructive. Dan Pink's definition of mastery and Carol Dweck's growth mindset both provide positive ways to boost a sense of achievement, enabling people to flourish and grow. When learning opportunities are not obvious, there are small things that managers can do to boost others' Self-esteem: asking people's opinions (so long as they are genuinely sought), trusting members of the team to carry out tasks that previously the manager would have done, giving praise and recognition for a job well done. They might be quite small actions but they can make a real difference to people.

The fifth domain in the SPACES model is 'Equity'. The ultimatum game is a good test of this (see Chapter 7). It is the game where one player is given money, and this player then proposes how to divide the money with a second player. The latter has the choice of accepting the offer, and both then get their share, or rejecting the offer, in which case both players get nothing. The majority of people reject offers of less than a third of the total amount (Camerer and Thaler, 1995). Receiving an unfair offer activates the anterior insula (Sanfey *et al*, 2003); a fair offer activates the brain's reward system. Tabibnia *et al* (2008) found that, compared with unfair offers of the same financial value, fair offers led to higher happiness ratings and activated regions in the reward network in the brain. A desire for justice is deeply rooted in us: what child hasn't at some point cried, 'But that's not fair!'? It goes back to survival again: in order to survive we need to get our fair share of food, warmth, etc. For adults, if there are going to be changes in the organization, then we need to feel that we will have as good a chance as the next person to be considered for the role that we want. When we feel the change processes in the organization are fair, then we are more likely to be able to focus on our work.

If we feel positive against all six domains, then we will feel very positive about work, but if an employee feels low on all six, they are in a very negative mindset. The question then for the manager is where can you begin to move people to a reward state? Although in some circumstances there might not be much we can do about, for example, Certainty, every manager can influence the quality of their relationships. There is always something that can be done to help to move the employee to the right of the diagram in Figure 9.1. As we know from earlier chapters, small changes can make a big difference to activating the threat and reward networks in our brains.

We all have our own preference in the domains – the one that really 'presses your button' and has the ability to make you feel very uncomfortable if you haven't got it and liberated if you have. This can change with time and circumstances. Pause for a moment and think about which one most matters to you. Then, remember projection bias (see Chapter 7), which is the bias where we assume other people see the world as we do. Don't assume your preference is everyone else's – ask people which matters to them. Every manager should talk to their team members about which domains are most important to them.

SPACES is a useful planning tool – from planning a meeting, conducting an appraisal through to planning transformational change. It keeps leaders constantly focused on what will keep people 'engaged' at work, emotionally and mentally.

# What can we do? Solutions and examples of what other leaders have done

All the models in this chapter are useful in helping to plan change both in advance and during the course of the actual changes. The steps suggested in the models require little or no financial investment: they focus on intrinsic motivators.

## Bridges' Transition Model

Bridges' Transition Model is a useful tool to help leaders identify where people are in the change process and what they can practically do to help people through. One UK government department that was going through downsizing used Bridges' model in workshops with managers. Managers found it particularly useful because it helped them to identify activities that both they and their teams could control. Under each of the three phases, Endings, Neutral Zone, and New Beginnings, managers shared ideas about what worked in their location. Here's a list of potential activities based on his model:

### Endings – what can you do?

- Listen.
- Don't be surprised by 'overreaction'.
- Help people let go of the past.
- Give people information.
- Mark the Endings symbolically.
- Treat the past with respect.
- Define what is over and what's not.

### Neutral zone

- Acknowledge that you are in this limbo zone.
- Create temporary systems, structures and roles.
- Protect people from failure.
- Use the Neutral Zone creatively/reframe it: it's an opportunity to try new ways of doing things.

### New beginnings

- Encourage new ways of doing things.
- Depict the future.
- Rebuild trust.
- Involve people in planning.
- Celebrate success.

## SPACES planning template – Hilary Scarlett

SPACES is useful for planning – from a phone call with a colleague, through to planning organizational change. It is helpful for planning in the moment, in the future and understanding retrospectively why someone might have behaved in the way they did. It also works on a personal basis – if you suddenly feel the threat state kick in, ask yourself what might have caused you to feel that way: has someone just undermined your Self-esteem, taken away some of your control, made you feel part of their outgroup? A useful conversation for managers to have with their teams is to describe the six domains and then ask people which one matters to them. This not only avoids projection bias, but is a very useful and positive conversation to have with people. It helps both the manager and team member understand what motivates them and it means that the manager is better placed from thereon to make decisions that help put the employee into a 'reward' state.

For planning, eg an appraisal conversation or a change intervention, it's useful to create a SPACES template. You are welcome to do this: please acknowledge Hilary Scarlett. Look at SPACES with Figure 9.1 (see p 168) in front of you as it provides a reminder of the impact of threat and reward states on people's brains. Below is a brief description of each of the six factors. Write down the six factors on the left-hand side of the page, and for each factor ask the question, 'What can I do to avoid a "threat" state and create a "reward" state in people's brains?'.

- Self-esteem: winning, feeling important relative to others, doing better than others, seniority, status, improving yourself, learning and developing, growing, mastering a skill, respected by others, sense of achievement, feeling valued and trusted.

- Purpose: making a difference, having a sense of direction, having meaning, hope, feeling that your contribution is useful, helping others, feeling needed.

- Autonomy: perception of having control over events or environment, free to determine how things are done, influence on decisions (even small ones), choice, not feeling constrained or micro-managed, feeling your voice will be heard.

- Certainty: predicting the future, knowing what is going to happen and when, clarity about responsibility, knowing what is expected of you, short-term goals.

- Equity: perceiving exchanges to be fair, transparency, the process is fair (especially important during times of change), the outcome will be fair, being treated with honesty and decency, fair play.

- Social Connection: feeling connected to other people, feeling part of an 'ingroup', feeling safe with others, feeling someone has your interests at heart and is interested in you, belonging and inclusion.

This should provide some useful, brain-friendly plans.

### Four enablers of employee engagement (Engage for Success)

The Engage for Success website is full of case studies and ideas on how to create an environment in which employees can feel engaged at work: **http://www.engageforsuccess.org/**

## Summary of key points from this chapter

- Change is unpredictable, and so more important than the change plan is the change planning process.
- Change is hard for the brain because change requires more energy than staying as we are.
- Change can cause a threat state and the impact of this is summarized in Figure 9.1.
- Leaders need to be aware of the small actions and words that can cause a threat or reward state in employees' brains.
- The five change/employee engagement models described in this chapter each provide a useful basis for planning.
- MacLeod's The Four Enablers of Engagement:
  - Visible leadership with a strong narrative
  - Engaging managers
  - Employee voice
  - Integrity
- Pink's *Drive*:
  - Autonomy
  - Mastery
  - Purpose
- Kotter's Eight Steps:
  - Establishing a sense of urgency

- Forming a powerful guiding coalition
- Creating a vision
- Communicating the vision
- Empowering others to act on the vision
- Planning for and creating short-term wins
- Consolidating improvements and producing still more change
- Institutionalizing new approaches
- Bridges' Transition Model:
  - Endings
  - Neutral Zone
  - New Beginnings
- SPACES – Hilary Scarlett:
  - Self-esteem
  - Purpose
  - Autonomy
  - Certainty
  - Equity
  - Social Connection
- In *Reflections and planning* below, there is a list of the 'headings' and relevant questions that might be useful in planning change based on the contents of this book.

# Reflections and planning

When planning change there are many areas that we need to take into account, including the following. You might want to make your own list, based on what you have read, and use this as a basis to plan.

1 Purpose.
2 Information and certainty.
3 Insight.
4 Autonomy.
5 Social connection.
6 Winning.

**7** Decision-making and bias.

**8** The inverted U of performance.

**9** Managing emotions.

## 1 Purpose

We know that having a sense of purpose is important, so:

- Is the organization clear about its overall purpose, why it does what it does?
- Is the purpose of the change clear?
- Is there an overarching or superordinate goal that both provides direction and gives people a shared purpose?
- What have you done to articulate why the organization needs to change?
- Is that articulation of the purpose clear and meaningful to people?
- Would using visuals and images make it more real for people?
- Do people know who will benefit from the changes and how?
- Have you connected people to the beneficiaries of the change and of their work?

## 2 Information and certainty

We know that the brain is a 'prediction machine'. Lack of information causes an error alert, and people are distracted and waste time and energy trying to fill the information gap, so:

- Do people feel well informed about the changes and how they affect them?
- Have you told people when and how they will receive information?
- Does the organization provide regular updates on how well it is doing against performance indicators and success criteria?
- Have you told people what is working well in terms of the changes?
- What more can you tell people?

## 3 Insight

We know that being told what to do creates a threat state in the brain, whereas having moments of insight is rewarding and leads to greater commitment to the changes, so: .

- Have you reviewed your communication plan to look at the balance between telling and enabling insight?
- How open have you been with the information that employees need in order to reach their own insights?
- What examples do you have where employees have been able to make their own connections/reach their own insights?
- Do leaders and managers have the skills and confidence to enable their people to reach insight (or do they tend to micro-manage and tell)?
- Do you allow time for employees to pause, reflect and make connections between the current situation and what needs to change?
- How strong are coaching skills in your organization?

## 4 Autonomy

We know that having control, or at least a perception of control, makes a significant difference to how employees perceive change and their levels of stress, so:

- Where are the opportunities for you to involve employees in the changes?
- What decisions might be better if you consulted employees?
- Where can you let go?
- Have you asked leaders where and how they are letting go of control? What examples can you give?
- Do employees feel they can influence what is happening in the organization at some level?
- What examples are there where involving employees has led to a better result?

## 5 Social connection

We know that much of what drives our behaviour both in and out of work is our need for social connection. We also know that social rejection reduces our ability to think, so:

- Are your leaders and managers aware of the impact of social connection at work?
- Do your leaders know the importance of showing warmth and then competence?
- Do all employees feel that they belong to a group within the organization?

- What language and tone do you use in your communications – is it inclusive?
- Is there a superordinate goal that provides a sense of shared purpose?
- Are there opportunities for more face-to-face communication?

## 6 Winning

We know that achieving goals generates dopamine and is rewarding to the brain, and that winning is a habit, so:

- What have you done recently to enable employees to feel like they are winning?
- Have you broken down the change into milestones that feel attainable?
- Do people feel they have short-term goals that they can reach?
- Do people get regular feedback on how the organization is progressing towards the change objectives?
- How do you celebrate success?
- When did you last ask people to recall past successes?
- Mindset: what have you done to encourage a growth mindset in your organization, such that people believe they can get better if they try?

## 7 Decision-making and bias

We know that making sound decisions during change is essential and that there is a trade-off between accuracy and speed. We also know that we are all prone to bias, particularly when we have power, so:

- How frequently have you scheduled decision-making meetings during the change?
- Does the timing of these meetings allow for the fact that decision-making can deplete our energy?
- What biases is your organization prone to?
- What biases are your leaders prone to?
- What interventions will you put in place to mitigate harmful biases?
- How will you ensure that both the process and outcomes of decisions are fair?
- How can you get different perspectives into your decision-making?

## 8 The inverted U of performance

We know from the inverted U of performance that some stress is good, but too much and the prefrontal cortex (where we do our considered thinking and planning) closes down, so:

- Where would employees say they are on the inverted U of performance?
- Is there a view that people need to 'do more with less' – is this sustainable?
- How many people are off work with stress-related illnesses?
- If people feel overwhelmed, how can you use autonomy and certainty to bring people back to the top of the inverted U?
- Do employees feel stretched and challenged?
- What can you do to bring novelty into the work of bored employees?
- What can you do to create a reward state in people's brains?
- How well do leaders and managers use extrinsic and intrinsic motivators?

## 9 Managing emotions

For the brain to perform at its best, we know that people need techniques to be aware of their emotions and to regulate them. They also need to find the right balance in terms of stress, glucose level, etc, so:

- Emotions are contagious: how good are your leaders at staying calm? Do they know the difference between suppressing emotions and managing them?
- What are you putting into place during the change to help people stay calm?
- Do people get enough sleep? What does the culture encourage?
- How do you help employees keep their glucose at an optimum level during the working day?
- When situations are difficult, how good are people at reframing, reinterpreting and using distancing techniques?
- Being always 'on' gets in the way of creativity and is stressful: is there a pressure on people to be always 'on' in your organization? Is it OK to have 'downtime'?

# References and further reading

Baumeister, R F and Leary, M R (1995) The need to belong: desire for interpersonal attachments as a fundamental human motivation, *Psychological Bulletin*, **117** (3), pp 497–529

Bernhardt, P C *et al* (1998) Testosterone changes during vicarious experiences of winning and losing among fans at sporting events, *Physiology and Behavior*, **65** (1), pp 59–62

Bridges, W (1991) *Managing Transitions*, Perseus Books, New York

Camerer, C and Thaler, R H (1995) Anomalies: ultimatums, dictators and manners, *The Journal of Economic Perspectives*, **9** (2), pp 209–19

Csikszentmihalyi, M (1990) *Finding Flow*, Basic Books, New York

Dweck, C (2007) *Mindset: The New Psychology of Success*, Random House, New York

Eisenberger, N I, Lieberman, M D and Williams, K D (2003) Does rejection hurt? An fMRI study of social exclusion, *Science*, **302** (5643), pp 290–92

Engage for success website [accessed 18 August 2015] [Online] http://www.engageforsuccess.org/

Grant, A (2014) *Give and Take*, Orion Publishing, London

Herry, C, Bach, D R, Esposito, F, Di Salle, F, Perrig, W J, Scheffler, K, Lüthi, A and Seifritz, E (2007) Processing of temporal unpredictability in human and animal amygdala, *Journal of Neuroscience*, **27** (22), pp 5958–66

Kotter, J P (1995) Leading change: why transformation efforts fail, *Harvard Business Review*, March–April

MacLeod, D and Clarke, N (2009) *Engaging for Success: Enhancing Performance Through Employee Engagement*, Department for Business, Innovation and Skills, UK

Marmot, M G (2004a) *Status Syndrome: How Social Standing Directly Affects Your Health and Life Expectancy*, Bloomsbury Publishing, London

Marmot, M (2004b) Status syndrome, *Significance*, **1** (4), pp 150–54

Pink, D (2009) *Drive: The Surprising Truth About What Motivates Us*, Canongate Books, Edinburgh

Redelmeier, D A and Singh, S M (2001a) Longevity of screenwriters who win an Academy Award: longitudinal study, *BMJ*, **323** (7327), pp 1491–96

Redelmeier, D A and Singh, S M (2001b) Survival in Academy Award-winning actors and actresses, *Annals of Internal Medicine*, **134** (10), pp 955–62

Robertson, I (2012) [accessed 16 August 2015] The Winner Effect at The Science Gallery [Online] https://www.youtube.com/watch?v=6JHWSP-W0FY

Sanfey, A G, Rilling, J K, Aronson, J A, Nystrom, L E and Cohen, J D (2003) The neural basis of economic decision-making in the ultimatum game, *Science*, **300**, pp 1755–58

Sapolsky, R M (2005) The influence of social hierarchy on primate health, *Science*, **308** (5722), pp 648–52

Tabibnia, G, Satpute, A B and Lieberman, M D (2008) The sunny side of fairness: preference for fairness activates reward circuitry (and disregarding unfairness activates self-control circuitry), *Psychological Science*, **19** (4), pp 339–47

Yerkes, R M and Dodson, J D (1908) The relation of strength of stimulus to rapidity of habit-formation, *Journal of Comparative Neurology and Psychology*, **18** (5), pp 459–482

# Changing behaviour

10

**M**ost jobs have an element of trying to change the behaviour of others: people who work in advertising and marketing are trying to get us to buy their products or buy them more often, dentists are trying to persuade us to look after our teeth and floss more frequently, teachers try to enthuse children with the pleasure of learning and the skills to do so, governments aim to get us to pay our taxes on time, social media companies are trying to persuade us to stay on their sites for longer. Within organizations, leaders want to inspire us to change our behaviour so that we can collaborate more, act as better advocates, innovate, focus on the customer, be open to learning, adapt to technology, work globally, be financially prudent, be flexible, be more inclusive, adopt the new values, become better coaches, be more engaging, improve our communication skills – the list goes on. Changing the way we do things is hard. It is hard to adopt new behaviours and it is particularly hard to let go of old ways of doing things. However, anyone who is involved in organizational change knows that, without behavioural change, there will be no real change in the organization.

There is no easy answer as to how we change our behaviour or other people's. Many, many things influence how we behave – upbringing, personality, past experience, education, expectation, personal circumstances, intrinsic motivation, external incentives and pressures, the behaviour of others, organizational culture, the need to survive. Much of what drives our behaviour is subconscious and automatic. That said, neuroscience and behavioural science are beginning to provide some insights as to how we can help ourselves to change behaviour and how to guide others to do the same.

Like neuroscience, behaviour change is an area that is still in its infancy and there is still a huge amount yet to be understood. It is already the subject of many books, articles and research papers, and companies are being set up to focus on this one area. In this book, it is just one chapter (although, in some ways, this whole book is about behaviour change); here we will take a look at why the brain finds learning new behaviours hard, the insights that neuroscience and behavioural science bring to how we can help change behaviour, and a few models that provide a useful guide.

# Our brains resist change

As we saw in Chapter 3, our brains are not designed to like change, particularly change that has been imposed upon us. They are 'prediction machines', which means that they don't want any nasty surprises; they want to conserve energy and they want to feel that they have some control over what is happening. Change presents a threat to all of these. Through various studies, Shaul Oreg (2003) of Cornell University identified four key factors that make people resistant to change, even when the proposed changes might be in both the organization's and the employees' interest; in other words, when resisting the changes might seem irrational. These four factors are:

1 routine-seeking;
2 emotional reaction to imposed change;
3 cognitive rigidity;
4 short-term focus.

Oreg's four factors cover behavioural, emotional and cognitive blockers to change. From what we know about the brain, these make sense.

## Routine-seeking

Our brains like habits and therefore like routine. Routine actions and behaviours take less effort and are less stressful to the brain; changes to these routines create uncertainty and present threats to our feelings of competence. We no doubt all have examples of doing things in a certain way – we know this might not be the best way, for example, to present the data, but it's the way we have done it for a long time, and it's the way we know how to, and so we carry on.

## Emotional reaction to imposed change

Change that is imposed upon us and not of our choosing takes away our sense of autonomy, which, as we know from Chapter 9, is important to the brain. Quite often imposed change feels like a threat to our self-esteem.

Working with employees who are being asked to change, I have seen people who feel affronted and undermined. They feel that being asked to change suggests that the way they did things before is now questioned and being judged as inferior or wrong. They perceive a lack of respect for the way they used to do things. Treating the past with respect is an important step in removing the threat response.

Some years ago I was working with a well-respected global airline. They decided to sell off part of the business to a little-known foreign company. Employees had been proud to say they worked for the airline and were not enthused about having to tell people in future that they would be working for a company no one had heard of. Both the acquiring company and the leaders of the acquired company were taken aback by the strength of this emotion but recognized that the emotion was heart-felt and should not be ignored.

Letting go of the past and feeling that something is being taken away from us is difficult. We looked at how sensitive the brain is to loss in Chapter 7. Even when change might be a success story, such as when a company is growing and expanding, this kind of change can present difficulties. Very often growth requires some loss of autonomy, such as a region that has to give up some of its control and local decision-making to be part of a larger, cohesive entity. In a tech company, employees who have recently joined recognize that the change might be easier for them because they have chosen to join a company that is evolving and growing, and their very arrival is contributing to the sense that things are now different. This transition might be harder for the incumbents – those who have been in the organization for years and did not actively choose to be in this situation. They have to learn to let go of autonomy, of the close-knit company it used to be.

## Cognitive rigidity

In his article, Oreg (2003) recognizes that various researchers have found that one trait which leads people to be more resistant to change is dogmatism or having a closed mind. Fear of, or lack of interest in, new ways of doing things might be one of the barriers to change. There might also be a fear of loss of face and self-esteem as we sense our lack of competence in learning new ways of doing things. Oreg suggests that people who tend to stick with the views they already have, and who therefore are less likely to change their minds, will be less willing or able to adapt to new situations, even if those new situations might be to their benefit.

## Short-term focus

As we saw in Chapter 7, one of our biases is temporal discounting: an example of this might be that a significant number of us would prefer £1 today rather than £3 tomorrow, but more of us would be happy to wait a day longer when asked if we would like £1 in a year's time or £3 in a year and a day's time. In other words, we don't mind our future selves waiting a day, but we don't want to wait right now.

The fact that CEOs are so often judged on their short-term results puts pressure on organizations to favour the short term over the longer term. This short-term focus might well be due to our brains still being so attuned to living on the savannah, where getting some food right now was more important than finding more food in six months' time. The more people focus on just the short term, the more they might lose sight of the longer-term goal and the reasons why imminent painful change might be worth it for the eventual gain. That said, human beings are among the few creatures who are capable of making short-term sacrifices to achieve the bigger goal. We will explore the impact of short-term and long-term goals on motivation later in this chapter.

# The science

In this section we will take a look at two types of learning that are relevant to behaviour change: habit learning and what is known as 'goal-directed' learning. Habit learning tends to be characterized by events that are linked closely together in time (eg when I walk into my house, I always take off my shoes), repetition (Hebbs' law: 'cells that fire together, wire together' – see Chapter 2), and a predictable outcome (taking my shoes off feels good and protects the floor from getting dirty). Context also affects habits (when I walk in the office door, I keep my shoes on): habits that are valid in one context are not necessarily valid in a different context. Much of what makes behaviour change hard is that a great deal of our behaviour and thinking is driven by habit. So, the first step is to understand a little more about habits.

## Habits

Habits are one type of behaviour and our brains like them. In large part, this is because habits allow us to perform routine actions without using too much of our conscious cognitive resources. Habits free up our brains to think about other, perhaps more interesting or challenging, things.

Think back to when you first learned to ride a bike or drive a car. Learning to cycle or drive is tiring and uses a great deal of mental energy. We have to concentrate hard. We need to develop hand–eye coordination that we don't yet have. When driving, we need to understand how our actions affect the car and, on top of all that, we have to watch other drivers, try to predict what they might do and make sure we don't damage their car or ours – oh and also try to avoid pedestrians (especially those who are staring at their smartphones and not looking where they are going) and not get flustered when we stall the car. Initially it's hard and requires our full concentration. We probably don't want music on in the car and we certainly cannot hold very much of a conversation. Once we are experienced, we can comfortably drive and talk (although not when something unexpected suddenly happens on the road – then driving needs our full attention again). Once we have learned to drive, it becomes a relatively easy task that requires less of our conscious attention and mental energy. It has become a habit. Habits are useful to the brain because it can push these tasks that we have had to learn from the conscious, effortful, System 2 (reflective system) into System 1 (reflexive). Relegating behaviours and actions into System 1 frees up System 2 for other things: see Chapter 7 for more on these two systems.

## Habits reduce stress

Reducing cognitive demand is one reason why habits are so useful to us. Research by Wood *et al* (2002) suggests that habits benefit us in other ways too. In their studies, students were asked to complete diaries every hour to report what they were doing. Between a third and a half of all behaviours listed were classified as habits. When performing habits, participants reported less stress and a greater sense of control. Any single behaviour that participants did not perform regularly increased their sense of stress.

## It's hard for the brain to 'unlearn' behaviours

Neuroplasticity: neural pathways establish and strengthen as we learn new things and apply that learning. The good news is that the new behaviour becomes less effortful. The bad news is that once we have established that neural pathway, the brain wants to keep using it. Trying to establish a new and different one is hard. It's similar to walking across a field of deep grass

– once the first person has crossed it, there's an indentation in the grass, the path becomes more established and easier for the tenth, twentieth and thirtieth person to use. It's much easier to use that path than to try a different route and wade through deep grass elsewhere. Neural pathways in the brain are similar. Once they are established, it is hard for the brain to 'unlearn' them and use a different route. They become remarkably fixed.

## Habits: cue–routine–reward (Duhigg, 2012)

In various studies of rats learning new habits and behaviours (Jog *et al*,1999; Howe *et al*, 2011), the pattern of neural activity can be seen to change as the new behaviour becomes more established. In her paper, 'Habits, rituals and the evaluative brain', neuroscientist Ann Graybiel (2008) describes this change in neural activity. When rats are put into a 'maze run' and are looking for chocolate, initially neural activity is heightened from the start to the end of the routine. Once they know where to look in the maze, ie they have formed a habit, neural activity becomes concentrated at the beginning of the task (the 'trigger') and at the end of the task (the 'reward'), with much reduced neural activity during the routine of going to get the chocolate. Brain activity changes as the brain learns a habit, and neural activities eventually settle into specific patterns. The rat thinks less and less during the routine part of the behaviour.

Whether habits are motor (that involve movement such as driving a car or typing on a keyboard) or cognitive thought patterns (such as always telling yourself: 'That team won't listen to me – they are never interested in what I have to say'), the basal ganglia play a key role in helping to form them. The basal ganglia are a group of structures found deep within the cerebral hemispheres, and include the substantia nigra, one of the few parts of the brain that create dopamine. Dopamine plays all sorts of roles and it is associated both with reward and habit. Initially, dopamine is created by the brain when we have a rewarding experience but, as we repeat the behaviour, this changes and our brains start to generate dopamine in anticipation of the reward. One of the challenges for drug addicts is that their brains start to produce dopamine in anticipation of the drug. Anticipation becomes more rewarding than the actual experience, so the experience is disappointing and leaves people craving more.

In his 2012 article for the *New York Times*, based on conversations with Ann Graybiel and other neuroscientists, author Charles Duhigg sets out the three stages of the habit loop: cue–routine–reward. He states that if we want to change our habits and behaviour, we need to be aware of each stage, analyse them, identify the cue, or trigger, and then change the habit. Understanding the components of habits makes them easier for us to control: this is about pulling habits out of the unconscious and back into the conscious.

Research by Stadler *et al* (2009) illustrates the point that understanding ourselves better and planning ahead helps us to build and sustain positive new behaviours. Gertraud Stadler and team recruited 256 German women aged 30–50. The women were invited to classes on the importance of exercise and then half of this group were invited to an additional class on the theories of behaviour change and in particular the importance of 'if/then' implementation plans (see more on this later in the chapter). They were asked to think about what they wanted to achieve in terms of physical exercise (get into better shape), what they would actually do (cycle to work, for example), potential obstacles and what might get in the way of the desired new behaviour (getting up too late) and then implementation questions that got them to plan for those obstacles. These included questions about when and where that obstacle might appear and what could they do to overcome it, and could they prevent the obstacle from occurring in the first place? Stadler *et al* found that those women who had been asked to think about how to form and stick to the new habit spent twice as much time exercising as the other group. The even better news is that these women were still acting on their good intentions when followed up 16 weeks later.

Duhigg's article goes on to tell the story of Febreze, Procter & Gamble's (P&G) air freshener. Initially the product was a disaster because P&G had introduced it by expecting people to create a new habit, eg hoping pet owners would use it to hide animal smells in the house. It was only when they observed one frequent user of Febreze that they realized it's hard to create new habits but it is much easier to attach a new habit to a routine that already exists. The customer they observed liked to use Febreze in each room as she cleaned it – squirting Febreze into the newly cleaned room became like a small reward at the end of her routine. As a result of this insight, Febreze became one of the top-selling products in the world. This story provides a useful insight – if we want employees to change their behaviour, we need to try to attach that new behaviour to something people already do.

We have taken a look at some of the research into habits; now let's look at some of the research into goal-based behaviour change.

## *Goals*

Goal-based learning is linked to dopamine. There is still a great deal that we don't yet know about this kind of learning. But we do know that because goal-based learning triggers dopamine, it feels rewarding to the brain, it's motivating and it's an efficient way to learn. There are many different types of goals: there are goals that we set for ourselves or goals that are set for us by others. We might have goals about what we need to do (I need to tidy up my desk) or about why (I need to be more organized). We might have goals that are about improving and learning ('get better') or goals about proving that I'm good already ('being good'), in other words, growth and fixed mindset goals – see more on this in Chapter 4. Our goals might be 'towards' or 'away' goals: goals that are about achieving something we want or goals that are about avoiding something unpleasant. Goals can be conscious or non-conscious. They can be individual or shared by a team. One additional caveat here: much neuroscience research is conducted in the lab (although this is changing as technology becomes more portable) and so tends to look at short-term goals; the neuroscience evidence to date is therefore limited.

## The importance of choice and autonomy

There has been a big shift in organizations in recent years, moving from top–down, control and command to more of a coaching-style of management. There are many reasons for this – recognition that people closer to the customer are better placed to make informed decisions about that customer, a reduction in deference in some societies, the rise of social media. Choice also makes a big difference to us. Put simply, we feel, 'If I've chosen it, I'm committed'. This is true in terms of objects but also in terms of behaviours. This has implications in many areas, including marketing, relationships and goal-setting. However, there have been questions about how enduring this commitment is. Research by Sharot *et al* (2012) demonstrates that once the choice is made, the commitment to that choice is long-lasting. Sharot *et al* asked research participants to make hypothetical choices about vacation destinations. In step one participants had to rank each destination on a scale of 1–6. In step two they were forced to make a choice between two destinations that appeared on the screen at the same time. Thirty per cent of these were an easy choice – they had already rated one higher than the other – but 70 per cent were hard choices – they had rated the two destinations the same and now had to make a choice of one over the other. In a control group, the computer made these choices for them. When people had to re-rate the destinations as they had in step one, they now rated the destinations they had chosen in step two more highly. Two and a half to three years later, the choosing of one destination over another still affected people's ratings of the destinations: they continued to prefer the destination they had actively chosen. In the control group, where the computer had made the choice for them, ratings for the destination chosen by the computer did not increase. This research demonstrates not only the impact of making a choice on how we feel about that chosen item, but also that our positive feelings endure. Making a decision and choosing something can lead to long-term change in preferences.

## Goals we have chosen – mPFC

To continue on this theme of choice, research by Lieberman (2007) shows that the distinction we feel between goals that we have chosen as opposed to goals that we have been given is also reflected in differing brain activity. When we choose our own goals, there's activation in the medial prefrontal cortex (mPFC) networks, whereas goals that we are given activate networks in the lateral PFC.

## Short-term or long-term goals?

When encouraging ourselves or others to change, what type of goals work best? Should we focus on the short term or the longer term? As we covered in Chapter 4, short-term, achievable goals can be useful if someone is having a tough day at work, as reaching a goal feels rewarding to the brain. But which is best and when? Research by Huang *et al* (2017) of Stanford Business School shows that short-term and long-term goals both have important roles to play but at different stages in the change process. Putting it at its simplest, in the early stages, we need short-term goals to get us going; as we get closer to the end goal, we need to focus on the goal itself. To use the metaphor of a road: as you embark on change, focus on the road just ahead of you; as you travel further, look up and focus on your destination. They also warn that if we don't get this right, the goals we set people can get in the way and impair performance. In their paper, 'Step by step: sub-goals as a source of motivation', they describe various experiments they have conducted. In one, the job required workers to collect market data over an eight-day period. Workers had to visit book stores and take pictures of certain books and upload those images. People received 'points' for each book uploaded and they received daily updates on how they were doing, but some were shown their efforts in relation to the long-term goal (target of 80 points), some were shown their results in relation to the daily target of 10 points. There was also a hybrid group who received updates relating to the daily goal for the first four days and then, for the last four days, got information as to how they were doing in terms of the overarching goal of 80 points. At the end of the eight days, the workers who were being told how they were doing in relation to the overarching goal uploaded 1,268 books; those working to the sub-goals throughout uploaded 1,392 books; but those in the hybrid condition, short-term goals first then focus on the longer-term goal, uploaded 1,906 books. Looking at the patterns of uploading, although those who were working to the overall goal throughout started as motivated as the others, their motivation was shown to drop by the end of the first day. Those working to the sub-goals throughout the eight days started off well but performance dropped during the latter days. The hybrid team's performance was similar to that of the short-term goals team in the first few days, but importantly, their motivation and performance stayed high during the last four days.

The work of Huang *et al* gives us useful insight into how to structure goals and the feedback we give on progress. Many organizations set daily, short-term goals, but this research demonstrates that we need to recognize when these help and when they undermine motivation. Huang *et al* also suggest that employees who are new to the organization might need more help in structuring their goals because they might have less confidence that the goals they are being set are achievable. Sub-goals might be all the more important in building the self-confidence of new hires.

Short- and long-term goals are processed in different parts of the brain. FMRI experiments by Stillman *et al* (2017) show that we process thoughts about future events or goals in the medial prefrontal cortex, whereas thoughts about short-term goals activate the precuneus, which is part of the superior parietal cortex.

## Goals and the impact of focusing on a higher purpose

Persuading us to behave in a more healthy way can be very difficult – in part because we often become defensive or go into denial when confronted with messages about what we should be doing. Research from the lab of Emily Falk at the Annenberg School for Communication and the Wharton Neuroscience Initiative found that a priming message or thought could lead people to be more open to messages about improving their health and, most importantly, had an impact on their behaviour. 220 sedentary people who were overweight or obese participated in the research (Kang *et al*, 2018). At the start of the research some groups were asked to think about matters beyond themselves, such as thinking about loved ones, things they deeply valued or that held important meaning to them. People often find these kind of thoughts to be intrinsically rewarding. In the fMRI scanner, these thoughts were seen to activate the ventral lateral PFC – a part of the brain associated with processing reward. A control group was asked to think about things that were not important to them. Then all the participants viewed plain-speaking health messages that encouraged them to be more active. The fMRI scanner revealed that those who had been primed with thoughts about things or people who mattered to them, on seeing the health messages, had more activation in the ventromedial PFC – a part of the brain that is activated by thoughts about ourselves and positive valuation. This suggests that the positive priming led to the participants being more receptive, and less defensive, to the health messages at a neural level. In the month that followed, the groups continued to get daily text messages that repeated the experiment in miniature. They also wore fitness trackers. Those who had thought about things that mattered to them were significantly more physically active in the following month.

This research has important findings for anyone who is trying to improve their own healthy behaviours or encouraging others to look after themselves better. There are also useful insights here for other kinds of organizational change: thinking beyond our own self-interest, connecting with deeper values and higher purpose, could remove some of the defensive threat response and act as an impetus to change.

Before we move on to look at some of the models we can apply to create and sustain behaviour change in the 'What can we do?' section, there are a few other useful points to be aware of: the role of shortcuts, anchors, priming, cognitive dissonance and the importance of mindset in achieving behaviour change. We will take a brief look at them here.

## *Our brains like shortcuts*

Shortcuts take effort out of decision-making and save us time. These decisions are taken in System 1, or the 'reflexive' system, without us even being conscious of them. On the whole these shortcuts are helpful as they provide a useful rule of thumb to guide our decisions, but sometimes they lead to flawed decision-making. Professor Robert Cialdini identified six shortcuts in his 2001 book *Influence*. The six are: reciprocation, commitment and consistency, social proof, liking and rapport, scarcity, and expert and authority; they are described at greater length in Chapter 7. All of these can influence how we behave. One shortcut that is particularly relevant in behaviour change is commitment and consistency. As Cialdini notes, once we have made a commitment in writing or to another person, we feel a pressure to be true to our word and to live up to what others expect of us. This is one of the reasons why getting people to put the commitment in writing or to have a 'buddy' who will check in whether we really have changed our behaviour and are sticking to it, can be so useful.

As Cialdini flags, when we are the ones who are seeking to change the behaviour of others, we need to tread with caution and consider the ethical implications. We need to 'nudge' for the good of the organization, the people who work in it and all other stakeholders. As you will see, some of the behaviour change models that we will look at in the 'What can we do?' section of this chapter are based on these six shortcuts.

## *Priming and anchoring*

Priming and anchoring are closely related. Priming is the effect when one stimulus influences the processing of a later stimulus. Anchoring is the bias whereby we rely too heavily on the first piece of information we are given. Some salespeople use anchoring, showing us a high-priced model of a car, for example, so that the not-quite-so-expensive model seems more reasonably priced. Bateson *et al* (2006) conducted an experiment in a workplace kitchen where there was coffee, tea and milk, and an honesty box for people to pay into for the cost of the drinks. Over a period of 10 weeks, Bateson and team changed the image above the honesty box: some weeks there was a picture of flowers, some weeks there were pictures of a pair of eyes. On average people paid 2.76 times more into the honesty box when there was an image of eyes rather than flowers. Although people were not actually being watched, they felt as though they were and this feeling led to more cooperative behaviour. The human perceptual system contains neurons that respond just to faces and eyes; Bateson *et al* suggest that perhaps these neurons were activated by the images of eyes in the kitchen. So, feeling that we are being looked at by others changes our behaviour.

It's interesting to note that priming is not just something we can do to others: we can prime ourselves by the images we choose to put on our desk, the music we choose to listen to or the words we read. Many elite athletes arrive on the tennis court or in the swimming pool listening to carefully chosen music to get them in the right mindset to compete at their best.

## *The role of cognitive dissonance in behaviour change*

As the research of Sharot *et al* (2012) shows, once we have made a choice we start to rate that chosen item more highly. Why do we do this? Festinger (1957) suggests that to reduce the discomfort of cognitive dissonance caused by difficult choices, we seek information that confirms that we have made the right choice.

## *Believing we can improve means we are more likely to*

As we saw in Chapter 4, having a growth mindset – believing that we can get better at things and having a willingness to learn from mistakes and setbacks – is an important belief to have if we want to improve. As we try to be more assertive in meetings or better coaches, we will make mistakes along the way. The narratives we tell ourselves about whether we are capable of changing or not are important.

# What can we do?

As 'The science' section suggests, changing behaviour can be difficult because so many different factors affect our behaviour and we are unaware of much of what causes us to do what we do. Competing priorities can also get in the way: people might want to be better at collaborating, but personal success is still what gets rewarded; a manager might want to let go and delegate but is worried about the consequences of a team member making a serious mistake. We want to get better at coaching and we know that it is good for long-term development of our people, but deep down we feel that we know best and it would be quicker and faster just to do the job ourselves.

There are three initial questions we need to ask:

1 What behaviour are we trying to change?
2 Is this something we can 'nudge' people into doing by playing on the shortcuts that our brains like to take, or is this a behaviour where people need to make a conscious decision to change?
3 How will we know if we have been successful?

If you are about to embark on trying to change behaviours in the organization, in the team or on a personal level, there are useful models that are worth looking at before you start. Let's take a look at four of them.

# Four useful models for behaviour change

## The COM-B model

The Centre for Behaviour Change at University College London in the UK has developed a model that raises useful questions about what we are trying to change and what might be getting in the way. Having reviewed the academic research, they developed the COM-B model (see Figure 10.1) (Michie *et al*, 2011). The B stands for 'behaviour' and, as Figure 10.1 shows, the model identifies three factors that influence whether people are able to change their behaviour. The C is for 'capability': do people have the psychological and physical ability to make the change? For instance, a business might be asking people to collaborate more globally, but has it given them the tools to share ideas and to work together? Psychologically, are people prepared to work and share ideas with people they have perhaps never or rarely met? One global travel company used storytelling, asking those people who had worked well together and who had been prepared to build on others' ideas to share those stories with other leaders. The participants could also share the difference it had made to their country's business in hard numbers, by not being shy about borrowing ideas from others. The O in the model is for 'opportunity': how easy is it for people to make the change? For instance, encouraging visitors and staff to use antibacterial gel in hospitals: a solution has been to put the gel dispensers everywhere, by each door, close to each patient, making it physically very easy to change the behaviour. The M stands for 'motivation': the mechanisms that activate or inhibit our

**Figure 10.1** COM-B model

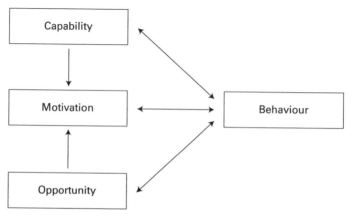

behaviour, whether we are aware of them or not. To change people's behaviour, we need to understand the world from their perspective and understand what is driving the behaviour.

According to this model, people need to have the capability, the opportunity *and* the motivation to change their behaviour. Often organizations tend to home in on one or two of these, but not all three. For example, a group of managers whom I met had been trained in storytelling but they didn't know why. They had been given the capability but not necessarily the motivation or opportunity.

To support behaviour change, the UCL Centre for Behaviour Change team has also developed the Behaviour Change Wheel (BCW). At the hub of the wheel is the COM-B model and this is surrounded by nine 'intervention functions'. The nine are: education, persuasion, incentivization (eg rewards and punishments), coercion, modelling (especially from leaders, acting as role models), restrictions (eg making people pay for plastic bags at supermarkets), enablement (providing the tools), environmental restructuring (such as putting recycling bins around the workplace) and training. Any intervention to change behaviour should provide at least one of these functions. Once we are clear about what the behaviour is that we are trying to change, which of the COM is lacking, we then need to reflect on the intervention. We need to identify the precise goal of the intervention in terms of what behaviour needs to change, in what way, how much and amongst whom.

The COM-B model has mainly been used to achieve changes in people's behaviour in health and medical environments, but it has been used in transport, finance and the built environment. For more information on how to use the COM-B model and the behaviour change wheel, visit the website: **www.ucl.ac.uk/behaviour-change**

## Cialdini – the science of persuasion

As a quick reminder from earlier in this chapter and Chapter 7, Cialdini identified six shortcuts that influence our behaviour; they are: reciprocation, commitment and consistency, social proof, liking and rapport, scarcity, expert and authority. You will see that Cialdini's work is reflected in the next two models – Mindspace and the work of the Behavioural Insights Team (BIT) in the EAST model.

## Mindspace

The creation of the Mindspace model (Dolan *et al*, 2011) acknowledged that much previous behaviour change work had been based on the belief that behaviour change was about changing minds through communication and incentives. However, evidence was emerging that the context and environment in which we work affects how we behave. Mindspace is a mnemonic and stands for:

- **M**essenger: we are heavily influenced by who communicates information to us.
- **I**ncentives: our responses to incentives are shaped by mental shortcuts, such as avoiding loss.
- **N**orms: we are strongly influenced by what others do.
- **D**efaults: we 'go with the flow' of pre-set options that save us from having to think.
- **S**alience: our attention is drawn to what is novel and seems relevant to us.
- **P**riming: our acts are often influenced by sub-conscious cues, such as the eyes above the honesty box.
- **A**ffect: emotional associations can powerfully shape our actions.
- **C**ommitments: we seek to be consistent with our public promises, and reciprocate acts.
- **E**go: we act in ways that make us feel better about ourselves.

This is a useful checklist when trying to shift behaviour. Starting with **Messenger**, we need to think about who the particular group of employees will listen to. A message from a leader we don't particularly like might lead to compliance, but we might change grudgingly, which reduces the likelihood of us sticking with that behaviour. On the other hand, if we like the leader or the message comes from someone who we perceive to be similar to us, we might be more willing to listen. Behavioural economics suggests that there are various factors we need to take into account if we are putting **Incentives** in place to change people's behaviour in the organization. As we know, people tend to value more highly things they receive in the short term than things they might gain in the long term. This is one of the reasons it can be so hard to get employees to save voluntarily for pensions. Getting people to imagine themselves or visualize themselves as old people has been shown to increase their likelihood to save.

**Norms** – as we have seen from the work of Cialdini, we are strongly influenced by what people like us tend to do. Applying this message has improved people's paying of tax (Halpern, 2015). **Defaults** and **Salience** – our brains are attracted to novelty and to things that are easy. Making healthy food in work canteens easily accessible has helped to change people's eating habits.

**Priming** affects us in many ways – words, images, smells, feelings. As we saw in Chapter 5, just the use of the word 'together' can change people's staying power. GCHQ in Chapter 12 describe how they thought carefully about the use of words when communicating about change. Footprints on the ground leading to a bin led to more people using that bin. Exposure to the scent of cleaning fluid caused more people to keep their table clean while eating in a canteen (Holland *et al*, 2005). Images of children on the pay packets of Indian workers led the workers to save more (Behavioural Change team). Images of people working together has led to people being more collaborative (Cialdini, 2001).

**Affect** – our emotions have a powerful effect on us, as we have seen throughout this book. Emotional responses are rapid and automatic: emotions kick in much more quickly than our reflective thought. The C of Mindspace is for **Commitment**, and Cialdini's research demonstrates this: if we make commitments in public, we feel a pressure to do as we promised. The final part of Mindspace is **Ego** – the need to feel okay about who we are and think well of ourselves, the desire not to let ourselves or others down.

## East – the Behavioural Insights Team

The Behavioural Insights Team (BIT), also known as the Nudge unit, was set up in 2010 in the UK and was the world's first government institution dedicated to applying behavioural sciences (Halpern, 2015). Its aim is to improve outcomes by better understanding what affects people's decisions and behaviour and to enable people to make 'better' choices for themselves. Cialdini and the *Nudge* authors Richard Thaler and Cass Sunstein were early influences on their work.

The BIT team has a four-part model to guide behaviour change, called East. These four factors make sense in light of what we examined in 'The science' sections of this and earlier chapters:

- Easy: people are much more likely to do something if it is easy to do. One example of this is more people contributing to their pensions when there is auto-enrolment.
- Attract: people are drawn to things that catch their attention and that are attractive and relevant to them. A good example of this is 10 times more doctors declaring their income when letters were directly targeted at them.
- Social: people are strongly influenced by what other people do. An example: people are seven times more likely to give to charity when they know a colleague has done so.

- Timely: interventions are more effective before habits have formed, or the behaviour has been disrupted in some way. Three times more workers choose healthy eating options a week ahead than on the day. Indian workers are more likely to save when they are paid twice a month rather than once a month (Quinn, 2018).

## *Changing habits and behaviour*

### Teach people about 'cue–routine–reward'

As we saw from 'The science' section, if we can understand how our habits work, we are in a stronger position to change them. So getting people to analyse what are the cue, routine and reward for them in their habits, and then attach a different routine to the cue, is a good place to start. Getting people to think about triggers and to make them obvious can be useful: perhaps having something on the desk, or in the meeting room, or on the factory wall that becomes the cue. It is easier to replace a habit than just stop, so once we have a cue, it's easier to change the routine that follows that cue than to ignore it.

### Link new behaviours to existing routines

Where can you attach a positive new behaviour to an existing activity? At a recent workshop, one participant talked about reflecting on what is good in life (see Chapter 11 for how this helps our brains). Every time she blow dries her hair, she reflects on what good things have happened that day – even very small things. In another case, one person said that every time they were about to open the fridge door at work, they would stop, pause and take a few deep breaths to calm down. For another person, turning on their laptop at work each day has become the moment when they think about their priorities for the working day (see Chapter 11).

### Priming

As mentioned earlier, we can prime ourselves. One person who wanted to get better at asking the team more questions put a small red sticker on her desk. That sticker became associated with the prompt to do less telling and ask more questions. You might also want to consider the images around the workplace, the music you play at the start of workshops, and asking the team to recall a time when they performed well despite the odds.

## New behaviours – give people choice

Earlier chapters in this book illustrate how important choice and autonomy are to the human brain, and this is no less true in the area of behaviour change. As we saw in 'The science' section, we process goals that we have chosen in a different part of our brain from goals that we are given by others, and our commitment to those choices remains stronger than to goals other people have given us.

## Encourage people to set their own goals

There are many reasons why organizations are enabling their leaders and managers to develop coaching skills. A manager who is coaching is not micro-managing; they are encouraging team members to think for themselves, to have moments of insight (which is rewarding to the brain) and to set their own goals.

## Create a climate where people ask for feedback

There cannot be many (any?!) workers who look forward to the appraisal process, and yet feedback on how we are doing is essential if we are to make progress. One of the ways some companies are trying to reduce defensiveness and the threat state of being given feedback is to encourage employees to ask for feedback. This way, they can take back some control of the process and can ask for feedback at a moment when they feel ready to listen and to deal with it. They can also prepare themselves with the questions they want to ask and the areas they want to explore.

## Goal-setting – articulating the new goal

As we have seen in earlier chapters, the very words and phrases we choose for the change we are trying to make have a significant impact on how we view that change. Our goals can be phrased as 'approach' goals or 'avoid' goals. Some people respond best to avoid goals, but many people respond better to positive ones. It's also important to make our goals precise and explicit; for instance, rather than, 'I'm not going to keep checking my e-mails so often', better to set a goal of, 'I will check my e-mails for 10 minutes and then again in one hour's time'. Goals that we can visualize are easier to attain: Chapter 8 touched on the point that the brain is better at storing information that is visual rather than verbal. So, where possible, get people to visualize what the new behaviour will look like in practice. The brain tends to favour things that come to mind easily, so make the desired behaviour easy to imagine and to remember.

## The why and how of goals

There are hierarchies within goals. Thinking about why we have a certain goal takes us up the goal hierarchy. Asking how we will achieve the goal takes us down the hierarchy. Goal hierarchies activate different parts of the brain – 'why?' is more abstract and activates the prefrontal cortex; 'how?' is more practical and activates the motor cortex (for more on hierarchy within the brain, see Matthew Botvinick's 2008 article). These are two different ways of thinking about goals. If people are struggling to achieve their behaviour change goals, it might be worth exploring whether they are getting stuck on why they have the goal, or how they can achieve it.

## Reward

What 'rewards' will people get for the new behaviour? Reward turns one-time behaviours into habits. Frequent rewards help establish the new behaviour. These rewards do not have to be financial; in fact, financial rewards can get in the way of sustained behaviour change as the behaviour might cease when the money stops. The social rewards set out in the SPACES model (Chapter 9) – Self-esteem, Purpose, Autonomy, Certainty, Equity and Social connection – are motivating and more sustainable than monetary rewards.

## Neuroplasticity

If people are finding it hard to learn a new behaviour or to form a new habit, remind them about neuroplasticity and that 'cells that fire together, wire together': keep practising and the brain will make the new behaviour easier and less effortful. It will move from System 2 (reflective) to System 1 (reflexive).

## Achieving our goals: focus on short-term targets, then the ultimate goal

The research by Huang *et al* (2017) demonstrates the importance of thinking carefully about when we should set short-term goals and when we should begin to shift people's attention to the endgame.

## What's your first, smallest step?

It's all very easy to go to workshops and seminars, leave with the best of intentions to change our behaviour or habits, and then do nothing. I'm sure we all have examples of this. One way to help people achieve those new goals is to ask them to plan the first tiny step they can take to make that change. It might be telling someone of their plan (see the next point), it might be blocking out 15 minutes in the diary to act on that change. Whatever it is, we just need to get started. Once we have made a start on that change, it is easier to keep going – we have some momentum.

## Implementation plans, or 'if/then rules'

There is a famous episode in the *Odyssey* where Ulysses knows that his ship is nearing the Sirens. He wants to hear their singing but knows that doing so will render him mad. He plans ahead: he will fill his men's ears with wax and asks them to tie him to the mast. Once within earshot of the Sirens, the men must ignore any of his commands to release him or to let him go nearer to the Sirens. This episode has led to these if/then contingency plans also being known as a 'Ulysses pact' – planning ahead for when we know we might be tempted to give in to unwanted behaviours or habits. If we really want people to change their behaviour, we need to get them to think about what might get in the way of their plans, and to be ready with a strategy that will keep them on track. The research conducted by Stadler *et al* (2009), referred to earlier in this chapter, demonstrates this well. For example, if there is one team member who is always being negative in meetings, being ready with how to deal with that negativity – perhaps asking other team members what they think, rather than rolling our eyes when off they go again – can be helpful. Getting the team or individual to think about the obstacles that might be in the way, and preparing for that moment, is an important part of achieving behaviour change.

## Make it as easy as possible for people to change

Changing behaviour is hard, so we need to make it as easy as possible for people to make the change and sustain it. There are plenty of examples of this – making contributing to a pension the default position, providing easy-to-use tools so that people can collaborate, providing healthy food in the company restaurant, providing sleep pods for naps.

## Priming

Priming can be a useful tool in the organization to encourage people to change their behaviours – the eyes above the honesty box in the kitchen, feet that direct people to the bin, the clean smell of a table, visuals of people working together.

## Feel accountable – find a buddy or make a public declaration

As Cialdini's research reflects, we like to be consistent and to be seen to be consistent by others. Asking people to commit in public to the behaviour change they want to make might help them to persevere in those difficult moments. It also means that, if other employees know what they are trying to achieve – for instance reducing biased decisions – they can give specific and appropriate feedback. If the behaviour change goal is one for the whole team, the closer the team is, the more likely they are to be committed to the collective goal.

## Get people to redirect their own narratives

To be able to change behaviour, people need to have positive narratives about themselves and they need the organization to provide positive narratives too. When trying to change behaviour, there will be moments when people fall back into old, unwanted patterns of behaviour. Having a growth mindset will be advantageous. As we saw in Chapter 4, people with a growth mindset are more likely to respond well to those moments of setback. They know they will make mistakes along the way, but they will be better equipped to deal with those challenges than people with a fixed mindset, who will be less confident about their ability to change.

# Summary of key points from this chapter

## *The science*

- Our brains like habits, in part because they free up our cognitive resources and they require less effort.
- Habits reduce stress and increase our sense of control.
- It is hard for the brain to 'unlearn' behaviours; once the brain has set down the neural pathway of a habit, it's hard to erase.
- Habit-based learning is linked to activity in the basal ganglia.
- There are three stages to a habit: cue–routine–reward. Once the habit is established and is routine, activity in the brain can be seen to shift towards different habit-related networks.
- Understanding and identifying the components of habit makes them easier to control and this can help us change our behaviour.

- It is hard to introduce a new habit or behaviour in others; it is much easier to link a new behaviour to a habit that already exists.
- Goal-based learning is linked to dopamine, which feels rewarding to the brain.
- Choosing our own goals leads to greater commitment to that goal.
- Goals that we have chosen as opposed to goals that we have been given activate different parts of the brain.
- The value of short-term or longer-term goals depends on how close we are to achieving the goal.
- Focusing on a higher purpose or things and people who deeply matter to us leads to different activity in the brain and can help people stick with their goals.
- Short- and long-term goals are processed in different parts of the brain.
- Our brains like shortcuts as they take the effort out of decision-making about how to behave.
- Priming can affect our behaviour without us being aware of it.
- We can prime ourselves.
- Believing we can improve means we are more likely to – ie, being in a growth mindset.

## What can we do?

- Behaviour change is difficult for many reasons, not least our own competing priorities.
- There are useful models.
- The COM-B model (UCL's Centre for Behaviour Change) sets out three conditions that must be met to achieve behaviour change: capability to behave in the new way, opportunity to make the change, and motivation. Organizations need to check that they are not focusing on just one, eg developing capability through training, but overlooking another, eg are managers working in an environment where they can display the 'new' behaviour? The Behaviour Change Wheel (BCW) sets out nine interventions to help achieve the behaviour change. These are: education, persuasion, incentivization, coercion, modelling, restrictions, enablement, environmental restructuring and training.
- Cialdini sets out six shortcuts that our brains like to take and that influence our behaviour without us necessarily being conscious of the

influence. The six are: reciprocation, commitment/consistency, social proof, liking/rapport, scarcity, expert/authority.

- Mindspace, developed by Dolan *et al* (2011) is also a useful checklist when we are trying to change behaviour. It stands for messenger, incentives, norms, defaults, salience, priming, affect, commitments and ego.
- East comes from the BIT team. If we want to change behaviour, we need to make it easy, attractive, social and timely.
- Habits – understand: cue–routine–reward.
- Link new behaviours to existing routines.
- It's hard to stop a habit; it's easier to replace it with a new one.
- In goal-setting we need to think about the language we use to describe the goal.
- If we want the behaviour change to stick, we need to give people choice over their goals – we are more committed to goals that we have chosen than to ones we have been given.
- Feedback is important in successful behaviour change, but most of us don't like receiving it. Encouraging employees to seek feedback and ask for it gives them more control over the process and therefore means they might be more open to listening and learning.
- We need to think about the progress reports people receive: research suggests people need to focus on short-term targets initially, and then as the goal gets closer, focus on how they are doing in relation to the overarching goal itself.
- People need to set small, achievable steps – just get started and get some momentum.
- Implementation plans are important; we need to think ahead to the times when we might be tempted to slip back into unhelpful habits or poor behaviours, developing if/then plans for those moments so that we are prepared for them.
- Make sure the work environment supports the desired behaviour – think about priming, and make the new behaviour easy for people to do.
- Have a buddy – someone to whom we are accountable.
- Growth mindset – people need to believe they can improve and that setbacks are a challenge, not a threat.
- The narratives people tell themselves and the narratives the organization tells are important for us to believe that we can change how we behave.

# Reflections and planning

Any organizational change requires some adjustment in how people behave. Changing behaviour, and sustaining that change, is hard. Before embarking on behaviour change, there are several questions we need to consider:

- What behaviour are you trying to change?

- Is this something you can 'nudge' people into subconsciously, or is this a behaviour that requires people in the organization to make a conscious decision to change?

- The COM-B model is particularly useful in analysing what might be getting in the way of behaviour change. What do people need or lack – capability, opportunity or motivation – to change?

- If people are resisting change, might one or more of the four factors identified by Oreg be the obstacle? Is the blocker due to people not wanting to let go of old routines, is it an emotional reaction such as fear of loss, are people closed-minded, or are they too focused on the short term?

- How will you know if you have been successful in helping people to change their behaviour? What will you measure?

Habits are difficult to shift, so:

- Do people understand the components of the habits they need to change? Have they identified the cue, routine and reward?

- How will you make people more aware of habits and how they work – can you give them some training to create greater self-awareness?

- It is hard to create new habits in isolation. What routines do people already have to which you or they can attach the new desired behaviour?

Goal-directed behaviour change can be supported by the organization, so:

- Choice is important to the brain and helps sustain commitment to the new behaviour. Is there scope for people to choose how they should change their behaviour?

- What targets will you set for people and what feedback will you give them on progress? Do they need short-term targets, and then switch to focusing on the ultimate goal?

- People will need feedback on how they are doing – how can you create a climate where people are open to feedback and are not defensive?

- Are people clear about why they need to change and in what way they need to change?

- Before embarking on behaviour change, would it be beneficial to get people to focus on thoughts about values, people or things that matter deeply to them?

- How are goals phrased – towards or away from? Which will work best?

- Are people's new behaviour goals specific enough that they can be measured?

The human brain likes to take the easy route, so we need to make the change as readily achievable as possible, so:

- What can you do to make the new behaviour easy for people? What can you change in the work environment? Are there any forms of priming that might help to influence their behaviour? Priming can take many different forms, including not just images, but smells and music too.

- We are deeply social creatures – who are the role models who need to be seen to be leading the way in changing behaviour and practices?

- Do people have the appropriate narratives in their heads and do they believe in their ability to change? What stories are they telling themselves about their ability to learn? Do they know about the growth mindset? How do they respond to setbacks and others' successes? People with fixed and growth mindsets respond differently to these (Chapter 4).

- Related to the point above, do they know about neuroplasticity and that you can 'teach an old dog new tricks' so long as the 'old dog' wants to learn? Not only that, but research suggests that learning difficult things is good for the brain and protective of our cognitive ability.

To succeed, behaviour change needs implementation plans, so:

- Planning to change behaviour is one thing, but have people thought about the first small step they will take to make that change? Gaining some momentum is key.

- Once started, sustaining the new behaviours is hard. How can you help people prepare for those difficult moments? Do people need a buddy or support from other people?

- Remember Ulysses. Have people made if/then plans so that they know what they will do in those moments when they are tempted to, for example, start micro-managing people again, consistently turn up to meetings five minutes late, not turn up to meetings at all, look for the problems rather than the solutions, resist new ideas, over-commit and under-deliver, or send e-mails over the weekend?

Rewards:

- How will people be rewarded for changing how they do things? Will the sense of achievement and self-esteem be enough?
- Rather than a monetary incentive, can you use one of the SPACES factors as a reward (see Chapter 9)?

# References and further reading

Bateson, M, Nettle, D and Roberts, G (2006) Cues of being watched enhance cooperation in real-world setting, *Biology Letters*, **2** (3), pp 412–14

Botvinick, M M (2008) Hierarchical models of behaviour and prefrontal function, *Trends in Cognitive Science*, **12** (5), pp 201–8

Cialdini, R B (2001) *Influence: Science and practice*, Allyn & Bacon, Boston

Dolan, P, Hallsworth, M, Halpern, D, Kind, D, Metcalfe, R and Vlaev, I (2011) Influencing behaviour: the mindspace way, *Journal of Economic Psychology*, **33**, pp 264–77

Duhigg, C (2012) How companies learn your secrets, *New York Times Magazine*, 16 February

Festinger, L (1957) *A Theory of Cognitive Dissonance*, Stanford University Press, Stanford, CA

Graybiel, A M (2008) Habits, rituals and the evaluative brain, *Annual Review of Neuroscience*, **31**, pp 359–87

Halpern, D (2015) *Inside the Nudge Unit*, Penguin Random House, London

Holland, R, Hendriks, M and Aarts, H (2005) Smells like clean spirit: nonconscious effects of scent on cognition and behaviour, *Psychological Science*, **16**, pp 689–93

Howe, M W, Atallah, H E, McCool, A, Gibson, D J and Graybiel, A M (2011) Habit learning is associated with major shifts in frequencies of oscillatory activity and synchronized spike firing in the striatum, *PNAS*, **108** (40), pp 16801–6

Huang, S, Jin, L and Zhang, Y (2017) Step by step: sub-goals as a source of motivation, *Organizational Behavior and Human Decision Processes*, **141**, pp 1–15

Jog, M S, Kubota, Y, Connolly, C I, Hillegaart, V and Graybiel, A M (1999) Building neural representations of habits, *Science*, **286**, pp 1745–49

Kang, Y, Cooper, N, Pandey, P, Scholz, C, O'Donnell, M B, Lieberman, M D, Taylor, S E, Strecher, V J, Dal Cin, S, Konrath, S, Polk, T A, Resnicow, K, An, L and

Falk, E B (2018) Effects of self-transcendence on neural responses to persuasive messages and health behaviour change, *PNAS*, **115** (40), pp 9974–79

Lieberman, M D (2007) Social cognitive neuroscience: a review of core processes, *Annual Review of Psychology*, **58**, pp 259–89

Michie, S, van Stralen, M M and West, R (2011) The behaviour change wheel: a new method for characterising and designing behaviour change interventions, *Implementation Science*, **6** (42)

Oreg, S (2003) Resistance to change: developing an individual differences measure, *Journal of Applied Psychology*, **88** (4), pp 680–93

Quinn, B (2018) How the 'nudge unit' pushed its way into the private sector, *Observer*, 11 November

Sharot, T, De Martino, B and Dolan, R J (2009) How choice reveals and shapes expected hedonic outcome, *Journal of Neuroscience*, **29** (12), pp 3760–65

Sharot, T, Fleming, S M, Xiaoyu, Y, Koster, R and Dolan, R J (2012) Is choice-induced preference change long lasting? *Psychological Science*, **23** (10), pp 1123–29

Stadler, G, Oettingen, G and Gollwitzer, P M (2009) Physical activity in women – effects of a self-regulation intervention, *American Journal of Preventative Medicine*, **36** (1)

Stillman, P E, Lee, H, Deng, X, Unnava, H R, Cunningham, W A and Fujita, K (2017) Neurological evidence for the role of construal level in future-directed thought, *Social Cognitive and Affective Neuroscience*, **12** (6), pp 937–47

Thaler, R H and Sunstein, C R (2008) *Nudge*, Penguin Books, London

Wood, W, Quinn, J M and Kashy, D A (2002) Habits in everyday life: thought, emotion and action, *Journal of Personality and Social Psychology*, **83** (6), pp 1281–97

# Planning the working day to maximize productivity

11

**W**e all have good days and bad days at work: what makes the difference? Why is it that some days, we can crack on with our work, think clearly, creatively and feel energized? Why is it that on some days we get off to a good start and then seem to lose direction and energy? Or perhaps it is the other way round, and we have a last spurt of energy and suddenly what seemed like an unproductive day turns into a constructive one? While this might seem to be a bit of a mystery, beyond our control, the good news is that we probably have more influence over these outcomes than we realize. A recurring theme in this book is that if we can understand our brains a little better, this understanding increases our say in what happens internally and externally too. We can set up our working days, be more self-aware and get ourselves back on track if we are having a difficult few hours. The aim of this chapter is to help us have more good days at work, or at least more good hours, and to make sure we are spending those good hours on the 'right stuff'.

## Some things to bear in mind

### Put on your oxygen mask first

Most of us are familiar with those safety videos on airlines that advise us that, in the case of an emergency, we should put on our own oxygen mask first, before helping others. We see that rather unusual image of the mother putting on her oxygen mask before helping her child. Surely parents are meant to, and want to, put their child first? But the message is obvious: you can't help others if you are struggling. The same is true for those leading and supporting change. We need to make sure we are coping. If we are feeling overwhelmed or burned out, we are not in the best state to support others. As we know from Chapter 6, emotions are contagious. If we are feeling frazzled and over-wrought, we will be spreading that stress to those who work around us. Far from helping, we will be hindering. Inadvertently, we will be causing others to feel more anxious and distracted and undermining their ability to work productively. But good news: positive emotions are also contagious and so we can be an influence for good too.

## Our brains have limited capacity

When thinking about our working day, most of us want to know how we can get our brains to think, focus, collaborate, articulate and innovate at their very best all day long. We want to be at our sharpest from the moment we start work until the moment we stop. But our brains can't do this: they have limited capacity. Why are our brains limited in this way? What limits them? What depletes our mental energy and are there ways in which we can restore it? In this chapter we will take a look at what fatigues our brains and what refreshes them. Some will seem obvious to us, but others might be more surprising.

## We probably have more control than we realize

To give our brains a good chance of being at their best when we are doing our most important work, we need to plan. As we will see (and have probably experienced), it's easy to waste our best hours on mundane administrative tasks, sitting in meetings we don't really need to be in, allowing ourselves to be distracted or surfing the internet. We need to plan what work we do and when. If we are aware of what fatigues the brain and what refreshes it, we can work with that knowledge and make better-informed decisions about how to use our day. Of course during the working day, unexpected requests are going to pop up and most of us are not in total control of how we spend our time. But, with more understanding of our brains, we can make better informed decisions about when to allow ourselves to be taken off course and then how to get ourselves back on track.

## The 21st-century work environment

The 21st century has seen great advances in technology and will continue to do so. One of the upsides of this is our improved ability to communicate with almost anyone, anywhere, in the workplace, at home, on the move. The downside is that we are almost always 'on' and feel that we are expected to respond to messages immediately. Some of the ways in which we are setting up the world of work are almost the opposite of what our brains need to be able to focus: the anxiety of having to find a desk to work at before the working day even starts, the noise in open-plan offices, e-mails and messages constantly popping up to distract us. One of the big challenges for our brains is fighting distraction and staying focused, and we are creating working environments that make this challenge even harder.

## The toll of mental fatigue and stress at work

As long ago as 2007, Ricci *et al* estimated that lost productivity due to fatigue cost employers in the United States more than $100 billion a year. According to the 2016 General Social Survey in the USA, almost 50 per cent of respondents say they are 'often or always exhausted at work'. In a 2017 survey in the USA (ComPsych), 59 per cent of workers said they had high levels of stress at work and felt out of control. In the same survey, 54 per cent of workers said they took 'stress breaks' at work to recover; but 35 per cent said their coping strategy was just to work harder. In Great Britain, 15.4 million working days were lost due to work-related stress, depression or anxiety in 2017/18; this represents 57 per cent of all working days lost (HSE, 2018).

## Mental fatigue – a useful warning sign?

We tend to assume that mental fatigue is intrinsically detrimental, but Boksem and Tops argue in their 2008 paper that fatigue might be a useful, adaptive signal to us. When our brains feel tired, they are sending us a message that perhaps we should stop what we are doing because the goal we are trying to achieve might not be worth the ongoing effort. It can be a nudge to change the goal or change what we are doing to achieve that goal. Mental fatigue might be telling us to take a rest. Managing energy levels is essential to achieving longer-term goals and to our well-being.

As with so much from applied neuroscience, small changes can have a big impact on our ability to perform and get the best out of our brains. There are many little things we can do to help ourselves have better days at work. As with earlier chapters, this one will take a look at some of the science behind what our brains need to work at their best (referring to, but without repeating, some of what we covered in Chapter 4, *Performing at our best during change*, and in Chapter 6, *Managing emotions during change*); the end of this chapter will provide some practical ideas on how to plan our working day and manage our mental energy.

# The science

## What causes mental fatigue?

We have all felt it: when the brain is fatigued it functions less well and we are much less able to pay attention and to concentrate. But what exactly causes brain fatigue? What leads us to feel mentally drained at the end of a working day?

As we saw in Chapter 2, our brains use up 20 per cent of our energy even though they represent just 2 per cent of our body weight. It would be easy to assume that when we are working on mentally taxing projects, the brain is burning even more calories. But that is not the case: this high rate of metabolism is remarkably constant no matter what we might be doing in terms of mental or physical activity (Raichle and Gusnard, 2002). Whether thinking hard or staring out of the window day-dreaming, our brains will be using up about the same amount of energy. When we are day-dreaming, the brain might not feel so busy but the default network is active, and even when we are asleep the brain is busy dreaming, storing information and making connections. Our brain is an organ and does not consume extra energy in the way that a hard-working muscle does. Many of the calories consumed by the brain are used just to keep it ticking over, so to speak, and for activities that are outside our consciousness.

It is still a bit of a mystery as to what exactly mental fatigue is and there are different hypotheses: is it what some call 'ego depletion'? Is it loss of motivation? Or something else? If you would like to read more about the hypotheses, a *Vox* article by Brian Resnick (2018) provides some analysis of the differing theories. For those who are interested in what is happening in terms of the brain, research by Lorist *et al* (2005) recognizes that we still don't fully understand what is going on, but mental fatigue is associated with a failure to get enough dopamine from the mid brain to the anterior cingulate cortex (ACC) and to the striatum.

That sense of brain fatigue can be caused by many things: working hard, boredom, lack of variety in our work, planning, making decisions, reining in our emotions, feeling anxious, staying focused, blocking out distractions, lengthy and difficult cognitive tasks (ie thinking hard!), lack of sleep, resisting chocolate. It should be noted that many of these can lead to brain fatigue but do not necessarily do so; for example, Boksem and Tops (2008) point out that working long hours does not necessarily lead to fatigue when we feel that there are positive rewards for putting in the effort – for example, when we feel our work is getting the recognition it deserves.

## Making decisions

Many years ago, I was listening to a North Korean defector being interviewed about both her dramatic and terrifying escape and her experience of living in the West. Were there any downsides to no longer living in North Korea, she was asked. Yes, she replied – the amount of choice and therefore decisions people in the West have to make. That wasn't such a problem in North Korea. Her response has stayed with me. At the time it surprised me that choice was seen as a negative but, on reflection, I understand. Recently speaking to a CEO, he talked about how his days are filled with having to make decisions. If people needed to speak to him it was almost always to ask him to make a decision. He found it exhausting. On a much more mundane level, I remember arriving in Nashville late one evening, after a long journey, tired and jet-lagged. With my co-worker, I ordered a drink and back came the well-meant questions: where did we want to have our drinks, inside or outside, which brand did we want, what size glass, ice or no ice, snacks or no snacks, what kind of snacks, large or small, hot or cold? 'Please, just bring me a drink,' my co-worker begged. They were all reasonable questions but our tired brains were struggling to deal with so much choice and so many decisions. Making decisions fatigues the brain.

As the North Korean defector recognized, having choice does not necessarily make life uniformly better. Vohs *et al* (2008) conducted a series of experiments exploring the impact of decision-making on our staying power, in line with the hotly debated ego depletion thesis. Some groups had to make choices and others did not. In one experiment one group of students had to make decisions about courses for their degrees and write down their choice. Another group was asked just to look at the curriculum but did not have to make a decision. All students were told that, in 15 minutes, they would have an important maths test to take and were told that the test was highly predictive of future success in the real world. They were also told that practising for 15 minutes beforehand has been shown to improve performance in the test and so the students were given practice maths questions. The researcher also told them that there were magazines and videos to look at, if they chose to. The researcher left the room and the students were observed by a research assistant, who did not know which students had just had to make decisions and which had not. In fact, the students did not have to take the maths test, but instead were debriefed and thanked. Those students who had not had to make a decision about future courses persevered with the practice questions for an average of 11 minutes 40 seconds. Those who had had to make a choice gave up revising after, on average, 8 minutes 39 seconds.

The latter group spent more time reading magazines and playing the video game. Vohs' work suggests that taking decisions reduces people's staying power and self-regulation. In another experiment, Vohs et al asked shoppers how many decisions they had had to make. They were then asked some simple maths problems. Vohs et al found that the more decisions the shoppers had made, the worse they performed on the simple maths task (the experimenters controlled for tiredness, gender, age and ethnicity). These findings are consistent with the hypothesis that making decisions fatigues the brain and impairs performance. This has important implications for those of us who work in organizations. How many decisions are we trying to make during the working day; having made those decisions, what impact has that had on our subsequent performance?

## Thinking hard – cognitive exhaustion

When people exert a great deal of cognitive effort, for example trying to complete unsolvable tasks, they make poorer decisions later on when faced with difficult but solvable problems. Sedek et al (1993) asked 36 high school students to look at various images (triangles, circles, some small, some large, some with stripes on, some without, etc) and were then asked whether each image contained a 'target feature': they had to reply 'yes' or 'no'. The problems were unsolvable as each image was paired half the time with confirmatory and half the time with contradictory information, so the answer was both 'yes' and 'no'. To add to the discomfort, participants were not told whether they were right or wrong, so they were left in uncertainty. The group was then asked to rate five films – they were asked to choose carefully because they would be given tickets to their chosen film. A second group of 36 students also participated in this ranking but had not had to go through the impossible decision exercise. Group one, who had participated in the decision exercise, spent less time than group two on the information provided about each film, suggesting they didn't want to think so hard, and were more likely to disregard their own criteria, ie they made poorer decisions. Cognitive exhaustion caused by the first task had undermined their ability to process information in the second task.

Centuries ago most workers would have been involved in manual labour and that work would have been limited by the length of daylight. For more and more people around the world, work is shifting from manual labour to cognitive work, and the end of daylight by no means marks the end of the working day. Work means that many of us have to engage in taxing cognitive work for lengthy periods. What impact might the resulting cognitive fatigue have on subsequent decisions we make? Mullette-Gillman *et al* (2015) have explored the impact of cognitive fatigue on economic decision-making. One group was given difficult memory tests over 60–90 minutes; the control group meanwhile watched relaxing videos for the same period of time. The research revealed that the first group was less consistent in the choices they made, suggesting that cognitive fatigue impairs the quality of our decision-making.

## Anxiety

Anxiety is often associated with a negative impact on performance. The good news is that, while this is often the case, it's not invariably so: anxiety can lead to us putting in more effort. The bad news is that this extra effort depletes our resources and makes us effective but less efficient (Eysenck *et al* 2007). By efficiency, Eysenck *et al* mean the amount of effort or resources we need to put in to achieve a task, with efficiency decreasing as we have to put in more of our resources to achieve the same level of performance. When we are anxious we have a reduced ability to control our attention and so are more easily distracted; the more anxious we are, the more prone we are to distraction. To add to our anxieties about anxiety, anxious thoughts consume the limited resources of working memory, meaning that less of our working memory is available for the task in hand. In addition to all this, anxiety leads our brains to pay more attention to threats – these might be 'internal' threats, such as negative thoughts and ideas in our heads, or negative events that are happening in the world around us. So, we need to be aware of what might cause us to be anxious during the working day, and where we can reduce causes of anxiety.

## Frazzled brains

In her 1998 article, 'The biology of being frazzled' in *Science* magazine, neuroscientist Amy Arnsten explores how our brains work differently when under stress. Some kinds of memory are altered – our memory for salient information such as the cause of the stress (eg a gruesome car accident) becomes heightened, while higher-order thinking (eg planning an important meeting) that is processed in the prefrontal cortex (PFC) deteriorates. When under acute stress our brains become distracted and disorganized (we are on the right hand of that inverted U of performance in Figure 4.2). The amygdala is activated and the PFC, the executive control centre of the brain, is suppressed. Memory is affected and retrieval of information is impaired. Brain recordings of rats by Maroun and Richter-Levin (2003) demonstrated exactly this: stress blocked plasticity/learning processes in both the hippocampus and the mPFC.

## Suppressing negative emotions

Being able to manage our emotions is a fundamental skill for success in life and at work. Imagine a day when you openly told everyone you came across what you really thought about them and their ideas. Suppression is a tactic we sometimes have to use in the short term to get us through the working day. There are times when we feel angry at what our manager has just said to us, or we are anxious about news we have just heard that could have negative consequences for our team, or we are embarrassed about just having tripped up in public – in all these cases we might feel it is best to keep our emotions to ourselves. As we saw in Chapter 6 from research by Butler *et al* (2003) and Richards and Gross (2006), suppression comes at a cost. In a sense it is a kind of multi-tasking: hiding the emotion at the same time as trying to carry on with our work or the meeting. Suppression is not simply the lack of expression of an emotion but requires our active effort to inhibit that emotion. We are trying to distract ourselves from what we really feel. Working in an environment where we constantly have to do this has an effect on us physically, emotionally and cognitively. One further interesting note on this: research by Butler *et al* (2007) suggests that cultural background might make a difference to this. Butler *et al* conducted a study and found that for Americans holding Western European values, habitual suppression was associated with self-protective goals and negative emotion. These negative outcomes were not so strong for people with more Asian values. In their 2007 paper, they suggest this difference might come about in part because Asian cultures encourage suppression in a broad range of situations and therefore suppression has different associations in the two cultures. Butler *et al* suggest that the group with more typically Asian values might associate suppression of emotions with more pro-social objectives, eg preserving relationships.

## Open-plan offices and blocking out distractions

More and more people work in open-plan offices. There are benefits to this – open-plan offices might require less space and therefore cost less; office walls are removed and so people see each other more and perhaps engage in conversations with co-workers with whom they might not usually speak. Or do they? Bernstein and Turban (2018) followed two companies that changed the architecture of their head offices, by removing walls and creating an open-plan work environment. Employees' interactions and behaviours were followed by use of unobtrusive portable technology that picked up on how much people were speaking and for how long, and from data on the servers that tracked the number of e-mails and messages. These were tracked over a number of weeks in the original working environment and in the new open-plan space. The tracking found that in both companies, once people were in an open-plan space, they communicated significantly less face-to-face, not just by a small amount, but 70 per cent less. They had replaced conversations with more e-mails and more instant messaging. This research is only based on two organizations, but it demonstrates that open-plan workplaces do not necessarily drive more face-to-face communication, and can achieve quite the opposite. Why might this be? Privacy matters. There is plenty of research that shows proximity drives social interaction, but human beings also have a strong need for privacy. Many people simply don't like open-plan offices. We need and like boundaries. If the office space does not provide them, it seems that we will protect our privacy in other ways – through e-mailing rather than speaking in front of others and through wearing protective headphones, to name just two tactics.

Having some privacy might also improve productivity. Bernstein *et al* (2018) found that intermittent rather than constant social contact produced the best performance amongst people involved in complex problem-solving. Open-plan offices can provide too much stimulation and distraction and so decrease productivity.

Evans and Johnson (2000) of Cornell University explored the impact of open-plan office noise. Female clerical workers were randomly assigned for three hours to one of two conditions: either a control condition that was quiet or a workplace of low-intensity noise, the latter designed to replicate an open-plan office. This research revealed two key findings. The first was that the women working in the equivalent of open-plan office noise made fewer adjustments to their posture than the women in the quieter environment. Not moving is a risk factor for musculoskeletal problems. Second, the group that worked in the noisier environment gave up more easily when asked to do puzzles (that were in fact unsolvable). The women did not report stress from working in the low-level noise – ie they were not conscious of feeling more stress – but the noise, without their knowing it, might have been having an impact on their motivation and mental energy. This could be due to the fact that in a noisier environment, we have to expend more mental energy blocking out distractions to focus on our work. This in turn is tiring and impairs our cognitive performance.

## Smartphones and devices

I ask people in workshops to turn their mobile phones off. Research now suggests that I should be asking them to turn them off and put the phone in another room. In a *Harvard Business Review* article in 2018, Duke *et al* describe two experiments they conducted amongst almost 800 volunteers. In one experiment people were asked to complete maths problems and memorize random letters. This tests how well people can keep track of task-relevant information while engaging in a difficult cognitive task. In the second, people saw a set of incomplete images and had to select the image that best completed the pattern.

There were three groups – all had to turn off phone sound and vibration alerts, then one group had to place their phones face down, another group were asked to put their phones into their bags or pockets, and a third group was asked to leave their phone in another room. Perhaps not surprisingly, the people who performed best were those whose phones were in another room, followed by those whose phones were in a bag or pocket. The worst performing group was the one whose phones were on the desk. Even though the phone alert signals were turned off, the people who performed worst were those who had their phone nearby. Even when our phones are on silent and face down, they are distracting and reduce our ability to concentrate and to think. Where is your phone as you are reading this?!

Media multi-tasking (consuming more than one item or stream of content at the same time) is becoming more common. Perhaps you know of people at work who insist that they can be on a conference call and deal with their messages at the same time, or participate in a workshop and respond to e-mails, or young people who insist that they can listen to music and do their homework and have lots of screens open at the same time?

Read on. A series of experiments by Ophir *et al* (2009) showed that the more frequent media multi-taskers actually performed worse at task-switching than people who are less frequent multi-taskers. They found it harder to filter out or to ignore information that was not relevant to the task. This research suggests that the very people who multi-task most are the worst at it. The impact of the internet on human cognition is understandably an area that many are interested in and is still the subject of much research and much debate (Loh and Kanai, 2015).

## Sleep

Sleep is so important to us and in so many different ways. There are now many studies that demonstrate the good things that happen in our brains while we are asleep, and the not-so-good consequences of being sleep deprived. We'll take a look at just a few.

In their 1997 article for *Nature*, Drew Dawson and Kathryn Reid underline the importance of getting a good night's sleep: they give the example of the increased number of accidents amongst shift workers who are short of sleep. In their study, one group of participants drank alcohol until their mean blood alcohol concentration reached 0.10 per cent, while another group were deprived of sleep from 8am until 12 noon the following day. At 30-minute intervals, both groups were given hand–eye coordination tasks to do. The performance of both groups declined significantly over time. After 17 hours of wakefulness, cognitive performance dropped to the equivalent of reaching 0.05 per cent alcohol concentration (around the allowed drink-driving limit in many countries); after 24 hours, performance dropped to the equivalent of 0.10 per cent.

We all know we feel more cranky and irritable when we are tired, but what goes on in our brains? What is happening at a neural level? Yoo *et al* (2007) took two groups, one of which was kept awake for 35 hours (two days and one night) and the other was allowed a night's sleep. The next day the groups were shown the same pictures, which ranged from neutral to emotional in content, while in an fMRI scanner. The amygdala – that part of the brain associated with processing emotions and particularly threats – showed 60 per cent more activation in the brains of those who were sleep-deprived compared with the control group. By comparison, the amygdala of those who had had a full night's sleep showed just a modest response to the emotional pictures. After a full night's sleep the prefrontal cortex (PFC) was able to inhibit the amygdala; lack of sleep impairs the PFC's ability to play its role.

Sleep can be extraordinarily beneficial in helping us to learn. In one study neuroscientists Walker *et al* (2002) set about teaching a group of right-handed people to type a number sequence on a keyboard with their left hand, the aim being to do it as fast and as accurately as possible. They then tested the group 12 hours later: group one had learned the sequence in the morning and so were tested again in the evening; group two had learned in the evening and were re-tested in the morning after a night's sleep. The group who were re-tested on the same day showed little improvement. Group two, who were re-tested after sleeping, showed a 20 per cent improvement in speed with no loss of accuracy. Group one was then tested again the following morning and this time, after a night's sleep, they showed a similar improvement to group two. This research applies to motor skills, but research by Ellenbogen *et al* (2007) shows that sleep also has a positive impact on our ability to make sense of, and connect, disparate but related information. In Ellenbogen's research, the group who had 'slept on' information made more accurate inferences in their decision-making than those who had not slept before the test. Interestingly, although the group who had slept performed much better, they did not report feeling more confident in their judgements than others. This suggests that sleep improves our performance without us being conscious of the boost.

Sleep, or perhaps to be more precise, dreaming, has an important role to play in creativity. There are many examples. Paul McCartney and Keith Richards both give examples of songs, *Yesterday* and *Satisfaction*, that came to them in their sleep. Mendeleev, the creator of the periodic table, said he saw the whole arrangement in his sleep. Thomas Edison is said to have been a keen daytime napper and would set himself up so that as he fell into the dreaming phase of sleep, ball bearings would drop from his hand into a metal pan and wake him up. He would then quickly capture any ideas that had emerged.

## Blood sugar levels

In Chapter 6, we saw the surprising and rather disturbing conclusions from research conducted by Danziger *et al* (2011), where chances of judges granting parole oscillated between 65 per cent and 0 per cent, depending on when the judge had last taken a break and had something to eat. Taking decisions takes its toll on the brain, and this research indicates that when we last took a break and ate affects our decision-making. A meta-analysis of 42 studies by Orquin and Kurzban (2016) found that when we are making decisions, low blood sugar levels lead us to be more impatient. This fits with Danziger's work. They also found that low blood sugar levels lead us to make more intuitive decisions rather than deliberate judgements, ie we use the fast System 1 rather than the slower, more reflective System 2 (except in decisions related to food, which makes sense given we have low blood sugar levels!).

## Boredom

As mentioned earlier, boredom is a stressor to the brain. A study by Wilson *et al* (2014) shows that some people would rather administer electric shocks to themselves than sit in a room by themselves with nothing to do. Yes, well let's move on.

## *What restores mental energy?*

On a positive note, there are many things, in addition to sleep and maintaining our blood sugar levels, that can restore our mental energy and function – taking a short rest, viewing scenes of nature, being in nature, achieving a goal, laughing, taking exercise, experiencing a positive mood, to name a few. Let's now take a look at what we can do to help ourselves have a better day at work.

# What can we do?

We need to make sure we spend our 'good' hours in the workplace in the best way and on our most important work. Based on the insights from neuroscience and other research, there are many things we can do to set up our working day to get the best out of our brains. Some of the scientific research described in 'The science' section above might have already triggered some thoughts. You might well already be acting on the suggestions in the following pages, or you might have had an inkling that they would help you take back more control over your working day. They are intended to be pragmatic. Open-plan offices and hot-desking are not how everyone would prefer to work, but for many of us they are a fact of life, and so we need to identify, within this way of working, what control we can wrest back. As with so much in behaviour change, the advice would be not to take on too much at once. Identify one or two things you might do differently, or do more or less of. Then it's about taking small steps and just getting on with them – not procrastinating. Once we have started, that gives us some momentum.

## *Managing our mental energy and planning the working day*

### 1 Prioritize your tasks for the day and write them down

The most important task of the day is to prioritize what we need to get done and to write this down. This is the very first thing we should do before we do anything else work-related. Some people choose to do this the night before as it helps to get the thoughts out of their heads and onto paper, and then they can sleep more easily. It might sound obvious, but you need to make sure that the list is achievable. In a recent workshop one participant recognized that she demoralizes herself before she even gets going in the morning as her list tends to be over 30 items; she knows she cannot possibly complete them all. We also need to identify which will make most demand on our brains – because they require either deep thought, creativity or attention to detail.

Write this list and set your agenda before others do it for you (see the next point on e-mails). Plan what you will do and when. Of course, during the working day many other unexpected demands will crop up. But as one leader said, having made his list of priorities, it meant that when he was approached during the day to get involved in other tasks, at least he could look at his list and decide whether he could or could not agree to be side-tracked. Prioritization provides our brains with a sense of greater certainty and a sense of control – both very important, as we have seen.

What are the most important tasks? Which ones need our brain to be absolutely at its sharpest or most creative? Have we got an important presentation this afternoon? If so, best not to use up all our mental energy this morning. Have we got some mundane administrative tasks? It's probably best to leave those until later in the day.

Come back to this list during the course of the day as you end a task. Pause and reflect, where is my time now best spent? Don't just drift unthinkingly to the next task – think consciously about how you are feeling, how much time you have, how your time is best spent now, and whether you need a 'quick win' to boost mental energy. It also has to be said there is great pleasure in striking through those tasks you have achieved.

### 2 Don't check your e-mails first thing

Many of us tend to check our e-mails and messages before we do anything else. In fact many people check them before they even get to the workplace. Once we start to read e-mails and requests from other people, we are allowing them to set our agenda for our day. Our aim should be to do point 1 first – make our own decisions about our most important work.

## 3 Don't respond to (all) e-mails first thing

It feels good to hit that 'send' button (probably because our brain gets a 'reward' signal from offloading a task), but even responding to e-mails uses up mental energy: we have to decide what to write, whom to copy in, what tone to take, etc. All these decisions are using up our thinking power, and if we have important documents to write or meetings to facilitate later in the day, we could have expended valuable mental energy on routine e-mails. So, we need to identify which e-mails merit an immediate response and which we can leave until later in the day.

## 4 Know what tires your mind

It's useful to note what kind of tasks we find tiring and then we can decide where to fit them into the day. I'll give a personal example: contracts from clients. I find them tiring because they require such an attention to detail, they are often very long, they are written in legalese rather than plain English and tend to be full of things we must or must not do and the dire financial consequences of not doing the right thing! Being aware of this enables me to think about at what point I will be best placed to take on checking the dreaded contract.

## 5 Understand how emotions affect your productivity

As we saw in Chapter 6, emotions have a profound effect on our brains. If you know you have a particularly challenging day at work, protecting yourself from strong, negative emotions is important. If there is a particularly emotional story in the news, choosing not to listen to that story can help maintain mental energy. Once we get drawn into the story or news updates are flashing on our mobile devices, the stories sap our energy. Having then to suppress those emotions as we return to trying to concentrate on documents, spreadsheets or project plans also reduces our energy. As we saw in 'The science' section, suppressing emotion reduces our cognitive ability and negatively affects our memories. Anxiety impairs our high-level, goal-directed attention: feeling anxious makes us more prone to being distracted.

## 6 Avoid small talk

This is one that has really helped me. I am frequently invited to speak at conferences. Often friendly people want to talk beforehand. But talking to people whom we don't know well can be mentally fatiguing. So, if you know you have a big presentation or an important meeting coming up, you might want to think about avoiding having to talk to people beforehand, and arrange to speak to them afterwards when the 'performance' is over and you are more relaxed.

## 7 Make decisions beforehand

How can you reduce the decision-making load on a busy day? As mentioned under point 1 above, some people like to write their 'to do' lists the night before. Making some decisions before a demanding day reduces the toll on the brain. One suggestion offered by Barack Obama, amongst others, is to remove some decision-making altogether. He was interviewed by *Vanity Fair* (Lewis, 2012):

> 'You'll see I wear only gray or blue suits,' he said. 'I'm trying to pare down decisions. I don't want to make decisions about what I'm eating or wearing. Because I have too many other decisions to make.'

Angela Merkel seems to share Obama's view. She tends to wear a 'uniform' of trousers and tunic, no doubt for similar reasons. She is a busy woman with bigger things to think about. Mark Zuckerberg of Facebook, likewise, is almost always seen in jeans and two types of grey t-shirt because he says he wants to save his mental energy for more important decisions at work. These might seem like small decisions, but reducing the workload on the brain is helpful. So, if you have a busy day coming up, take what decisions you can at least the day before.

## 8 Block out thinking time

As stated earlier, the way we are setting up the world of work is almost the opposite of what our brains need to focus. For most organizations, the quality of people's thinking is the difference between success and failure. People need good thinking time and this is very difficult to achieve in a workplace where people can approach you at any moment and disrupt your train of thought, or a message can pop up on a device. Where we can, we need to block out and protect thinking time. This might be once a week, but even once a month is a start. This should be time when we turn off our smartphones and leave them in another room, we tell co-workers that we need to be left in peace, and we remove ourselves from being easily accessible by others – either by finding a quiet place in the office or at home, or perhaps by going for a walk. Most of us are employed for the quality of our thinking and our ability to solve problems or have new ideas. We need to give our brains the chance to do this. As we saw in Chapter 4, those 'aha!' moments tend to come to us when we have a quiet brain.

## 9 Get thoughts down onto paper – frees up working memory

Our working memory is limited. We have a limit of about three to five items that we can hold in mind (Cowan, 2010). By writing things down we are reducing the cognitive load on our brains, and this in turn makes us less vulnerable to distraction.

## 10 Take a break

We saw from 'The science' section that mental fatigue has an impact on our cognitive function, so it makes sense that taking a break will help. How frequent and how long that break should be depends on what we are doing. Research by Henning *et al* (2010) suggests that even three-minute breaks every hour – when people who have been working at computers do a bit of stretching – can help productivity and lead them to feel more comfortable physically.

## 11 Don't multi-task

What we might perceive as multi-tasking is just task-switching in the brain. It's an inefficient way of working since we actually lose time as our brain switches between tasks (APA, 2006). Media multi-tasking is becoming more common but, as Ophir *et al*'s (2009) research showed, those people who are frequent media multi-taskers are more prone to distraction as they have greater difficulty in filtering out irrelevant information. So we need to stop fooling ourselves: if we really want to do good work, we need to focus on one task at a time.

## 12 Set intentions, choose your filters: prime yourself

Work can be quite unpredictable, especially when we are going through change. Lack of variety and boredom can also be stressors. The circumstances in which we work can be hard to control, but we have more control than we probably realize over how we respond to the work environment. How our brains perceive the world is influenced by a wide variety of factors, including past experience, our expectations and personal circumstances. Confirmation or 'myside' bias means that we tend to see what we are looking for. If we expect the people we are about to meet to be difficult and obstreperous, that is probably what we will experience. We have set our filters to look for the signs that prove that we are right. As we have seen throughout this book, being in a threat state leads us to look at the world through a filter of threat, to be oversensitive to threats that exist and even see them where they don't. But, if we pause before the start of each working day or before each meeting, think about what outcome we are after, and set our intentions, we can change both our experience of that meeting and our impact on that meeting. Emotions are contagious, both positive and negative, and we can choose to set a positive, kinder tone. What meetings have you got coming up over the next couple of days? How can you set your filters, what good things will you look out for, to make it more likely that the meeting is constructive?

## 13 Learning to stay in a 'reward/toward' state

This one is closely related to setting intentions. Knowing what activates the reward state in our brains gives us more choice and control over getting into, and staying in, this more positive mindset. Having a tough moment? Ways of making it less tough might include reminding ourselves of a time when we have done well, a moment we feel proud of (to the brain, visualizing that moment is almost the same as experiencing it again), thinking about what we do like about our jobs rather than all the things that are annoying us right now (and perhaps even remembering, if it is true, that working in this organization is a choice). Setting ourselves a short-term goal that we can achieve – completing it feels rewarding to the brain and puts it into a better place to take on the next challenge.

## *Mind–body connections*

There is an inextricable link between the mind and the body and each can influence the other.

## 1 Maintain blood sugar levels

Once we have heard about that research by Danziger *et al* (2011), it's hard to forget. We need to reflect on how this affects us at work. If we are interviewing people, are we being as fair to the person we see late in the morning, when we have not eaten for a while or had a break, as the person we saw first thing? Some people skip breakfast or lunch. What kind of impact is that having on their decisions? We need to make sure we are eating regularly and in a way that keeps our blood sugar levels as steady as they can be. Hydration is also important for the brain.

## 2 Learning to calm the brain on demand

Mindfulness – the skill of learning to switch off the inner narrative in our heads in a non-judgemental way – continues to be one of the most researched areas in terms of reducing stress at work. Chapter 6 covers some of the wide range of research into the impact of practising meditation or mindfulness. Some people ask how mindfulness compares with relaxation techniques. Research by Jain *et al* (2007), using a randomized control trial, found that just one month of mindfulness meditation has positive effects on reducing distractive and ruminative thoughts and on enhancing positive states of mind, compared with both a waitlist control group and a group who undertook relaxation training. That said, there are some caveats: research by Hafenbrack and Vohs (2018) suggests that mindfulness brings about calm and acceptance, but can reduce motivation in some cases.

## 3 Running from meeting to meeting? Stop: take a breath

In the midst of a busy day, we might not always have time to practise mindfulness or meditation for any length of time, but just stopping – and pausing – to take a few deep breaths can help. Another technique that can be practised anytime and anywhere is called 'open monitoring'. Put simply, this is about remaining aware and 'in the moment' – just noticing and observing thoughts, sensations and feelings as you experience them, without judging them. This is about being aware of thoughts but not being lost in thought.

Taking a deep breath sends a signal to the brain that we are relaxed and helps to calm us down. Many of us have to race from meeting to meeting, and each of those meetings might require us to be in a different mode: at some we have got to be upbeat and energize those around us; at another we might have to explore why something went badly wrong on a project; at another we might be pitching for new work. Between each of these meetings we need to stop, pause, take a breath, review what we want to achieve and set our intentions for how we will behave.

## 4 Laugh

Laughter is a great de-stressor of the brain. Chapter 4 referred to the work of Oswald *et al* (2009) at Warwick University, where the IQ of people who had a good laugh at a comedy DVD increased by around 12 per cent. Research conducted by Berk *et al* (1989) showed that stress hormones such as cortisol reduced more quickly in a group that laughed than in a control group.

## 5 Sleep/take a nap

All animals sleep. We need sleep because it is beneficial for both cognitive and neural health. As we saw in 'The science' section, sleep is important for processing thoughts and consolidating memories. It helps us to regulate our emotions. We all know after a bad night's sleep how easy it is to get cranky and how hard it is to feel positive. If you have missed out on a good night's sleep, neuroscientist Jessica Payne (2015) suggests taking a 20-minute nap (but no longer) will be beneficial for refreshing the brain. She also suggests that you might want to drink some caffeine just before you nap. When it's time to finish the nap, the caffeine should kick in and help you focus.

We need to create work cultures where it is okay to nap: we need to get rid of the belief that it's all about effort and activity. We are biological beings, not robots. We're not all fortunate enough to work in organizations that recognize that sleep and napping are good for productivity. Google is well known for having sleep pods in the office, but most organizations do not. If taking a nap is not yet acceptable in the workplace, try to meditate for a few minutes, or go for a mindful walk.

## 6 Physical exercise

There's plenty of evidence that taking short moderate exercise like going for a jog or a brisk walk has immediate positive benefits on our mental functioning and mood and makes us feel more energetic (McMorris and Hale, 2012). Research by Legrand et al (2018) suggests that 15 minutes of short brisk exercise leaves us feeling more energized and, in turn, benefits our attentional control and mental speed – more so than 15 minutes of relaxation. So, making sure that we take some short sharp exercise during the day is good for the body and good for the ability to think.

One idea many people are putting into practice, especially if it is a one-to-one meeting, is to go for walking meetings. Even the busiest of people have time for this.

## 7 Productivity, not activity

One final point we all need to keep in mind – don't mistake activity for productivity! Don't take on too many of these suggestions – it's much better to do one or two things well.

## *What does a healthy mind need each day? ACCESS*

Just as we are recommended to eat a certain amount of vegetables and fruit each day to maintain a healthy body, so we should think about what a healthy brain needs each day. ACCESS summarizes six things our brains need each day:

1 Acknowledge – taking a few moments to acknowledge what is good in our lives helps us to reframe positively.

2 Calm – developing techniques to calm the brain on demand is essential for dealing with a busy work environment.

3 Challenge – challenging the brain to learn new things is important and might be cognitively protective (Feldman-Barrett, 2017). Learn something difficult.

4 Exercise – important for the body and for the brain, as mentioned earlier in this chapter.

5 Social – spending time with friends and family: that sense of belonging is important.

6 Sleep – this chapter has illustrated how important sleep is for learning, consolidating memories and emotional regulation.

It is useful for each of us to reflect on this. Are we doing enough of these activities each day? How about team members – do we need to encourage them to spend more time with friends and family, get some brisk exercise or just have a bit of laughter and fun?

# Summary of key points from this chapter

## *The science*

- Our brains have a limited capacity and cannot work at their very best all day long. To get the best out of our brains and to make the most of our mental energy, we need to plan.

- Brains use up around 20 per cent of our energy, and this does not vary that much, whether we are thinking hard, or daydreaming.

- There are differing theories as to what depletes our mental energy and the cause is still not clear – is it 'ego depletion'? Loss of motivation? Mental fatigue has been associated with a failure to get enough dopamine to the ACC and striatum.

- Feeling mentally fatigued might be advantageous in that it's a sign to consider whether what we are doing is worth the effort.

- Many things deplete our mental energy – making decisions, cognitive exhaustion, anxiety; working long hours does not necessarily lead to fatigue if, for example, we feel there are positive rewards for the effort we are putting in.

- Cognitive exhaustion can lead to poorer decision-making.

- Anxiety reduces our ability to control our attention.

- Our brains work differently under stress – our memory for salient information can improve, our ability to recall other information is impaired.

- Suppression of emotions comes at a cost, but this might be affected by different cultural values.

- Open-plan offices can lead to less face-to-face communication as people try to protect their privacy.

- The noise of an open-plan office can have a negative impact on our motivation and mental energy, without us being conscious of the impact.

- Devices such as smartphones, when they are within sight, impair our thinking, even when they are turned to silent.

- People who are frequent media multi-taskers have been shown to be worse at task-switching, making their brains inefficient.

- Sleep is beneficial in many ways, including improving emotional control, learning, making sense of information, and creativity.

- Blood sugar levels affect our decision-making.

- Boredom is a stressor to the brain.

- Many things can help restore our mental energy.

## *What can we do?*

- The most important task of the day is to write down what we have got to get done and to prioritize.

- We should prioritize our activities before we check e-mails.

- We need to be careful about not being lured into responding to all e-mails – this could deplete our mental energy.
- We need to understand what tires the mind.
- We need to reduce decision-making where we can.
- If we need good thinking time, we need to block out time.
- Get thoughts out of your head and onto paper or devices.
- We need to avoid multi-tasking.
- Set our intentions for the day and for each meeting constructively. We need to stay mindful of what will help keep our brains in an open, constructive state. We have more control than we think.
- Try to keep our blood sugar levels steady.
- Learn to calm the brain on demand.
- Laugh: it's a great de-stressor.
- Get enough sleep/take a nap.
- Take exercise – walking meetings are a useful practice.
- Focus on being productive, not busy.
- For a healthy mind, use the ACCESS model: Acknowledge the good things in life; learn to Calm the brain on demand; Challenge the brain – learning is good for us; take Exercise; be Social – spend time with friends and family; get enough Sleep.

## Reflections and planning

We have more control over our mental energy and productivity than we probably realize, so:

- Do you know when your mental energy tends to be at its highest or lowest? Make a note to become more aware of this and plan your working day accordingly.
- Do you know what depletes and what restores your mental energy?
- How can you avoid or reduce the things that tend to eat into your mental energy and do more of the things that restore it?
- What behaviours or habits do you want to change to improve your working day? What's the first small step you will take?

# References and further reading

American Psychological Association (APA) (2006) [accessed 2 March 2019] Multitasking: switching costs [Online] www.apa.org/research/action/multitask

Arnsten, A (1998) The biology of being frazzled, *Science*, **280** (5370), pp 1711–12

Berk, L S, Tan, S A, Fry, W F and Napier, B J (1989) Neuroendocrine and stress hormone changes during mirthful laughter, *American Journal of the Medical Sciences*, **298** (6), pp 390–96

Bernstein, E S and Turban, S (2018) The impact of the 'open' workspace on human collaboration, *Philosophical Transactions of the Royal Society B*, https://doi.org/10.1098/rstb.2017.0239

Bernstein, E, Shore, J and Lazer, D (2018) How intermittent breaks in interaction improve collective intelligence, *PNAS*, **115** (35), pp 8734–39

Boksem, M A S and Tops, M (2008) Mental fatigue: costs and benefits, *Brain Research Reviews*, **59**, pp 125–39

Butler, E A, Egloff, B, Wihelm, F H, Smith, N C, Erickson, E A and Gross, J J (2003) The social consequences of expressive emotion, *Emotion*, **3** (1), pp 48–67

Butler, E A, Lee, T L and Gross, J J (2007) Emotion regulation and culture: are the social consequences of emotion suppression culture-specific? *Emotion*, **7** (1), pp 30–48

ComPsych Survey (2017) [accessed 13 January 2019] StressPulseSM survey [Online] www.compsych.com/

Cowan, N (2010) The magical mystery four: how is working memory capacity limited and why? *Current Directions in Psychological Science*, **19** (1), pp 51–57

Danziger, S, Levav, J and Avnaim-Pesso, L (2011) Extraneous factors in judicial decisions, *PNAS*, **108** (17), pp 6889–92

Dawson, D and Reid, K (1997) Fatigue, alcohol and performance impairment, *Nature*, **388** (235), 17 July

Duke, K, Ward, A, Gneezy, A and Bos, M (2018) [accessed 2 March 2019] Having your smartphone nearby takes a toll on your thinking, *Harvard Business Review* [Online] https://hbr.org/2018/03/having-your-smartphone-nearby-takes-a-toll-on-your-thinking

Ellenbogen, J M, Hu, P T, Payne, J D, Titone, D and Walker, M P (2007) Human relational memory requires time and sleep, *PNAS*, **104** (18), pp 7723–28

Evans, G W and Johnson, D (2000) Stress and open-office noise, *Journal of Applied Psychology*, **85** (5), pp 799–83

Eysenck, M W, Santos, R, Derakshan, N and Calvo, M G (2007) Anxiety and cognitive performance: attentional control theory, *Emotion*, **7** (2), pp 336–53

Feldman Barrett, L (2017) [accessed 13 January 2019] How 'superagers' stay sharp in their later years, *Observer*, 30 April [Online] www.theguardian.com/

science/2017/apr/30/work-on-your-ageing-brain-superagers-mental-excercise-lisa-feldman-barrett

General Social Survey (GSS) (2016) [accessed 13 January 2019] www.gss.norc.org

Hafenbrack, A C and Vohs, K D (2018) Mindfulness meditation impairs task motivation but not performance, *Organizational Behavior and Human Decision Processes*, **147**, pp 1–15

Health and Safety Executive (2018) [accessed 13 January 2019] Work-related stress, depression or anxiety statistics in Great Britain, 2018 [Online] www.hse.gov.uk/statistics

Henning, R A, Jacques, P, Kissel, G V, Sullivan, A B and Alteras-Webb, S M (2010) Frequent short rest breaks from computer work: effects on productivity and well-being at two field sites, *Ergonomics*, **40** (1), p 1997

Jain, S, Shapiro, S L, Swanick, S, Roesch, S C, Mills, P J, Bell, I and Schwartz, G E R (2007) A randomized controlled trial of mindfulness meditation versus relaxation training: effects on distress, positive states of mind, rumination and distraction, *Annals of Behavioral Medicine*, **33** (1), pp 11–21

Legrand, F D, Albinet, C, Canivet, A, Gierski, F, Morrone, I and Besch-Richard, C (2018) Brief aerobic exercise immediately enhances visual attentional control and perceptual speed: testing the mediating role of feelings of energy, *Acta Psychologica*, **191**, pp 25–31

Lewis, M (2012) [accessed 5 January 2019] Obama's way, *Vanity Fair* [Online] www.vanityfair.com/news/2012/10/michael-lewis-profile-barack-obama

Loh, K K and Kanai, R (2015) How has the internet reshaped human cognition, *Neuroscientist*, pp 1–15

Lorist, M M, Boksem, M A S and Ridderininkhof, K R (2005) Impaired cognitive control and reduced cingulate activity during mental fatigue, *Cognitive Brain Research*, **24**, pp 199–205

Maroun, M and Richter-Levin, G (2003) Exposure to acute stress blocks the induction of long-term potentiation of the amygdala-prefrontal cortex pathway in vivo, *Journal of Neuroscience*, **23** (11), pp 4406–09

McMorris, T and Hale, B J (2012) Differential effects of differing intensities of acute exercise on speed and accuracy of cognition: a meta-analytical investigation, *Brain and Cognition*, **80** (3), pp 338–51

Mullette-Gillman, O A, Leong, R L F and Kurnianingsih, Y A (2015) Cognitive fatigue destabilizes economic decision-making preferences and strategies, *PloS ONE*, **10** (7), e0132022

Ophir, E, Nass, C and Wagner, A D (2009) Cognitive control in media multitaskers, *PNAS*, **106** (37), pp 15583–87

Orquin, J L and Kurzban, R (2016) A meta-analysis of blood glucose effects on human decision-making, *Psychological Bulletin*, **142** (5), pp 546–67

Oswald, A J, Proto, E and Sgroi, D (2009) *Happiness and productivity*, IZA Discussion Paper 4645, IZA, Bonn, Germany

Payne, J (2015) [accessed 4 March 2019] Your brain on sleep [Online] www.nd.edu/stories/your-brain-on-sleep/

Raichle, M E and Gusnard, D A (2002) Appraising the brain's energy budget, *PNAS*, **99** (16), pp 10237–39

Resnick, B (2018) [accessed 5 January 2019] Why your desk job is so damn exhausting, *Vox.com* [Online] www.vox.com/science-and-health/2018/9/5/17818170/work-fatigue-exhaustion-psychology

Ricci, J A, Chee, E, Lorandeau, A L and Berger, J (2007) Fatigue in the U.S. workforce: prevalence and implications for lost productive work time, *Journal of Occupational and Environmental Medicine*, **49** (1), pp 1–10

Richards, J M and Gross, J J (2006) Personality and emotional memory: how regulating emotion impairs memory for emotional events, *Journal of Research in Personality*, **40**, pp 631–51

Sedek, G, Kofta, M and Tyszka, T (1993) Effects of uncontrollability on subsequent decision-making: testing the cognitive exhaustion hypothesis, *Journal of Personality and Social Psychology*, **65** (6), pp 1270–81

Vohs, K D, Baumeister, R F, Schmeichel, B J, Twenge, J M, Nelson, N M and Tice, D M (2008) Making choices impairs subsequent self-control: a limited-resource account of decision-making, self-regulation and active initiative, *Journal of Personality and Social Psychology*, **94** (5), pp 883–98

Walker, M P, Brakefield, T, Morgan, A, Hobson, J A and Stickgold, R (2002) Practice with sleep makes perfect: sleep-dependent motor skill learning, *Neuron*, **35** (1), pp 205–11

Wilson, T D, Reinard, D A, Westgate, E C, Gilbert, D T, Ellerbeck, N, Hahn, C, Brown, C L and Shaked, A (2014) Just think – the challenges of the disengaged mind, *Science*, **345** (6192), pp 75–77

Yoo, S S, Gujar, N, Hu, P, Jolesz, F A and Walker, M P (2007) The human emotional brain without sleep – a prefrontal amygdala disconnect, *Current Biology*, **17** (20), pp R877–78

# Applying neuroscience in the organization

12

There is a great deal of interest in neuroscience and its applications but it is still early days: the field is only just emerging from the lab and into the workplace. When talking to people about the practical application of neuroscience, many have asked, 'Where are the case studies and the examples of leaders applying neuroscience to their work?' 'What difference has it made?' In the first edition of this book, I described some research I conducted with my colleague Michael Pounsford where we asked groups of leaders from four large but different organizations – Lloyds Banking Group, Department for Business, Energy and Industrial Strategy, BAE Systems, and Orbit Housing Group – to work with us over time. They learned about neuroscience and then applied some of this learning to their work. For this edition of the book, we have spoken again to some of those participants to see what impact learning about neuroscience was having several years on – more on this later in the chapter.

Before revisiting that research group, let's take a look at three organizations which have been learning about and applying neuroscience to their work. At the time of writing, these organizations are at different stages. Two have just embarked on encouraging leaders and managers to learn more about the brain; the other has now been using the insights for several years. With those in the earlier stages, we will focus more on the process of sharing the learning; with the organization that has been applying the learning for longer, we will look at how they have been applying the learning and what outcomes they have seen.

# Tier-one investment bank

The technology team of this tier-one investment bank is in the earlier stages of learning about and applying insights from neuroscience. They decided to run a pilot amongst a group of 20 vice-presidents (VPs), the aim being to see whether learning about the brain would enable each of the VPs to make some small shifts in behaviour to improve both their own performance and that of their teams.

## What we did

There were four key stages to our approach:

1 **Set-up:** creating the right environment where people wanted to learn. We talked to some of the VPs and their team members to understand what would motivate them to come to the workshops. This proved useful, particularly in how to communicate about the learning and what people would gain from it.

2 **Measurement:** identifying how we would know if the learning had had an impact. We based our measurement on the Kirkpatrick framework (Table 12.1): we looked at reactions to the workshops and follow-up,

**Table 12.1**   Measurement of the sessions based on the Kirkpatrick framework (Kirkpatrick and Kirkpatrick, 2009)

|  | **Description** | **Areas to evaluate** |
| --- | --- | --- |
| **Reaction** | What did people think about the sessions and follow-up? | Usefulness of the sessions and what people have learned |
| **Learning** | What did people learn? Feedback collected following the sessions | Changes introduced, knowledge applied, insights or awareness generated that has stayed with the participants, impact of change |
| **Behaviour** | How has behaviour changed as a result? | Observable changes in people's approach and actions, and style of managing others |
| **Results** | What impact has there been? | Operating performance measures Measures set up by participants |

we measured what people had learned, we asked both the VPs and their direct reports what behaviour change they had seen as a consequence, and with what results.

3  **Deliver three workshops, face-to-face:** see below for our approach to how we designed the sessions and the outline content for each.

4  **Embed learning between workshops:** we stayed in contact with the VPs between each workshop to provide 'nudges' and reminders. Between each session, VPs committed to trying out something new as a result of what they had learned in the previous session.

## Harness the power of your brain – our approach

Before we started the sessions, we agreed a few ground rules, a key one being to switch off devices during the session to enable people to concentrate and give their full attention.

Workshops were designed to maximize learning. This meant presentations were kept short, and there was plenty of time for people to discuss and reflect what the learning meant to them. We encouraged people to take hand-written notes, to think about who they might share the learning with (this helps learning to stick) and used a mixture of methods – films, quiz questions, and short experiential exercises. Talking about the threat state is all very well, but it makes much more of an impression on people when they experience it and see what immediate impact it has on their ability to think and be creative, and the experience usually ends in laughter. We also allowed time in each session for people to reach their own insights about what the learning meant to them and how it might help them in the workplace or outside it. This is when the real learning is done – when people apply what they have just heard to their own experience. We made sure there were plenty of refreshments in the room so that people could keep a good level of energy and hydration – important both for the brain and for learning.

## Content of the modules

Workshop 1 covered some basics about the brain: if we can understand a little about our brains, we can work with that knowledge rather than in ignorance of or despite it. This initial section provided the foundation for the rest of the workshops. We then looked at the social brain and our need for social connection at work. We also explored how working as part of a diverse team can sharpen up our critical thinking.

Workshop 2 focused on growth mindset (see Chapter 4), small actions that make a big difference to performance and the SPACES model (Chapter 9).

Workshop 3 explored techniques to stay calm under pressure (Chapter 6), what fatigues the brain, and how to plan the working day to get the best out of our brains (Chapter 11).

In each of the three workshops, people identified a goal for the coming month and set out what they would do and the first step they would take to reach that goal. At the start of Workshops 2 and 3, we asked what people had been doing and with what results. We also asked people to provide feedback and evaluation at the end of each session so that we could see what was working and what had really struck home. The workshops were spread four to six weeks apart, and between each session we shared films or articles that we thought might be of interest and that would keep the learning 'front of mind'.

## *What people did*

At the time of writing, these workshops have only recently been completed, but participants have reported back that they have already made some small changes in behaviour and are beginning to see positive consequences both for themselves and in others.

Many of the VPs manage people who are located on another continent and whom they rarely see face-to-face therefore. Several said that following the first workshop they had become aware of the importance of social connection on people's productivity. They had therefore changed the way in which they started some of their phone calls to people. Rather than rushing straight in to talk about the business issue, they were aware of the importance of taking some time to talk about the person and what they had been doing at the weekend, for example. Others said they were trying to send fewer e-mails to people who are based in the same building and walk over to talk to them instead. Some had decided to get their team members to sit in amongst the internal client team to improve relationships and understanding. Both the clients and direct reports had found this useful.

One said he had decided to micro-manage less. He said not only had the team 'stepped up' and got more involved in making decisions, but this had also led the team to consult more with other people, to get their input before making a decision. So, there had been two 'wins' here.

Others had been struck by how easily the threat response is triggered in our brains. They said they had tried to be mindful of this and thought carefully about how they interacted with team members, in some cases asking more questions (as we know from Chapter 11, being asked a question can be rewarding to the brain). Some had used the SPACES model with their direct reports to have a discussion about motivation and how the leader could best work with them. Many talked about taking time to give praise and positive feedback where it was due, and said they very quickly began to see a positive response in the direct reports.

Working in technology can be very demanding: the team has to respond extremely quickly to any issues. It is therefore not surprising that many were interested in techniques that would help them to stay calm when under pressure. Several said they wanted to try practising mindfulness, others wanted to think about how they organized their working day so that they tackled the more demanding tasks when their brains were not fatigued.

Growth mindset and appreciative inquiry were two concepts that many put into practice immediately after the workshops. They were keen to create a culture that is one of learning and so began to ask more questions: what have we learned from our mistakes and what can we learn from what we have done well?

## Evaluation of the three workshops

The bar chart in Figure 12.1 shows the positive reactions and shifts in learning over the three workshops – all measures moved in a positive direction. The three strongest positive shifts over the three sessions were:

- 'I know how to organize my day to get the best out of my brain.'
- 'I understand the impact of the "threat state" on people's ability to focus at work.'
- 'I understand the impact of feeling excluded at work on people's ability to think.'

Reactions to learning about neuroscience from the technology team were extremely positive over the three sessions, as is shown in Figure 12.2.

**Figure 12.1** Tier-one investment bank – harness the power of your brain

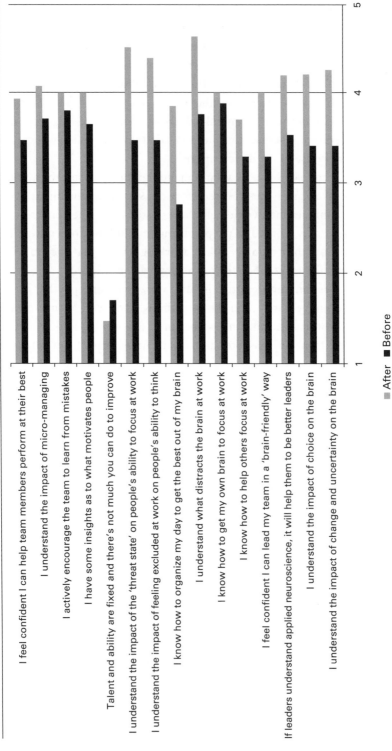

■ After  ■ Before

**Scale of 1 to 5:**
1 – Strongly disagree / 2 – Disagree / 3 – Neither agree nor disagree / 4 – Agree / 5 – Strongly agree

**Figure 12.2** Tier-one investment bank – overall evaluation of the sessions

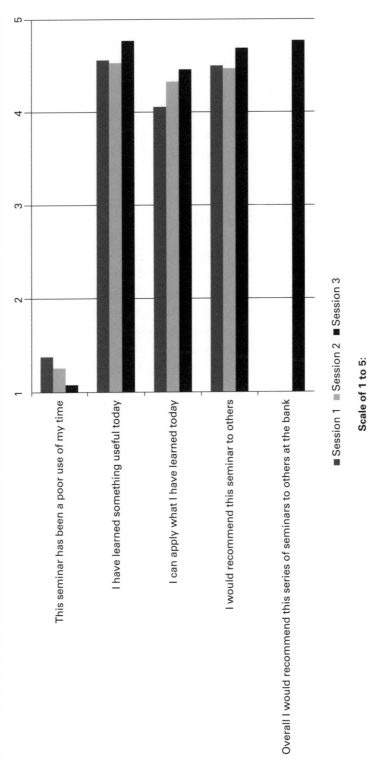

■ Session 1  ■ Session 2  ■ Session 3

**Scale of 1 to 5:**

1 – Strongly disagree / 2 – Disagree / 3 – Neither agree nor disagree / 4 – Agree / 5 – Strongly agree

At the end of the sessions, participants commented that learning about neuroscience had been an 'eye-opener', and had given them a 'completely different perspective'. They liked the 'very different perspective compared with standard corporate training' and they said that all levels of management in the bank should learn about the brain.

At the time of writing, the technology team is now sharing the learning more widely, in particular offering 'Harness the power of your brain' sessions to new recruits and their managers to help the recent graduates settle in faster.

# Lloyds Banking Group

Lloyds Banking Group is a leading UK-based financial services group providing a wide range of banking and financial services for both personal and commercial customers. The group's main business activities include retail, commercial and corporate banking, as well as insurance, pensions and investment provision. The group operates the largest digital bank in the UK as well as the largest branch network.

The part of the bank called Enterprise Transformation was created to support and protect the group and its customers, and to respond to the competition brought by technology and 'fintech' leaders in financial services. The team uses data to develop innovative products and technologies that enable them to deliver both a leading customer experience and to transform the way people work in Lloyds Banking Group.

So, why was the Enterprise Transformation team interested in using applied neuroscience? Emmanuelle Hazem, Culture Lead for the team, explains:

> Our world is changing exponentially and it is hard for people to adjust to changes we are seeing all around us – globalization, climate change, technology and the way we work, to name just a few. All these have real impacts on how we live our lives and make decisions. It is often said that 'change is the only constant', and within Enterprise Transformation we are at the heart of the transformation of the bank, introducing significant changes, for instance, in our ways of working and our approaches to solving problems. In this context, being able to lead people through change becomes vitally important. We were interested in scientific evidence and meaningful insights from applied neuroscience – in particular how our brain works, the impacts the 21st-century workplace has on it – how leaders could use this knowledge in day-to-day interactions with their teams.

## An experiment

The Enterprise Transformation team decided to design an 'experiment' for the leadership team. The experiment tested the following hypothesis:

1 People who understand some key facts about what the brain needs will be better equipped and more confident to lead people through change.

2 They will understand what helps both their own brains and those of their teams to work at their best.

3 By understanding what causes distraction and stress to the brain, they will be better placed to remove or reduce some of those causes of stress.

The participants in the experiment were 20 leaders from Enterprise Transformation, led by Gill Wylie, Director and Senior Executive. The experiment consisted of two workshops, one in June and one in October, with the leaders receiving 'nudge' e-mails in the months in between. Follow-up phone calls were also offered. Enterprise Transformation measured the knowledge and attitudes of the leaders before the first workshop and after each session. The measurement showed very positive shifts, for example:

'I understand the impact of change and uncertainty on the brain'
Before: 43 per cent strongly agree/agree
After: 100 per cent strongly agree/agree

'I understand the impact of choice on the brain'
Before: 43 per cent strongly agree/agree
After: 100 per cent strongly agree/agree

'I feel confident that I can help team members perform at their best'
Before: 64 per cent strongly agree/agree
After: 100 per cent strongly agree/agree

'As a result of today, I feel better equipped to help people perform at their best at work'
100 per cent agree (67 per cent strongly agree; 33 per cent agree)

At the end of the first workshop each leader chose a goal that they would work on between the workshops. These are very busy people and so they were encouraged to choose something small that they could start on right away. Some identified the need to create more thinking time so that they could be less reactive, others wanted to start to practise mindfulness. Some decided that they would note down their priorities for the day before checking e-mails. Others decided to apply the SPACES model (see Chapter 9), and were particularly interested in strengthening social connection and reminding people of their purpose and of the benefits their work brings to others.

The second workshop focused on staying calm under pressure, building resilience and managing mental energy. In particular we explored a demand so many leaders face – how do you run from meeting to meeting, where in one meeting you might be welcoming new team members, in another you might be having a conversation about poor performance, and another might be dealing with a significant problem? How do leaders do all this, be 'mentally present', and hit the right tone in each one? How do people reach the end of the day not feeling mentally depleted? One decision the team took was not to allow back-to-back meetings, but to start meetings at, say, 10 minutes after the hour to give people a chance to pause, breathe, think about their aims for the next meeting and then set their intentions.

Jo Brown, Transition Lead for Enterprise Transformation and Group Design & Data, comments:

> 'Insights from neuroscience have enabled me to link our transformation strategy with our people in a more human way. We all know that change can be difficult but understanding, for example, that social exclusion lights up the same part of the brain as physical pain has led me to see patterns in teams that I knew existed but could never explain before. Now, rather than applying a one-size-fits-all lens to helping people through change, I seek to understand what is making a person feel as they do and, using one of the many tools at my disposal, to help.

## Government Communications Headquarters (GCHQ)

Government Communications Headquarters (GCHQ), one of the UK's three intelligence and security organizations, has not only taken a deep interest in applied neuroscience, but has put its insights into practice. Based on the first edition of this book, change agents, organization development and leadership development professionals have been putting the learning into action to improve leadership, increase resilience during change and reduce the 'threat state' at work. In this section, three people, who we will call Katharine, Charlotte and Lynne, share how they have been using applied neuroscience to help the organization through change. Since I first met some of the team two years ago, they have been very busy applying the learning in many varied ways.

## Book club

Based on the popular idea of book clubs, GCHQ has formed its own, focused on work-related books. One of the first things the GCHQ team did was to suggest the book club read the first edition of this book – the aim being to learn and to discuss what they could take from it.

## Workshops

Following a workshop I ran with the team, the group took some of the ideas and incorporated them into 1.5-hour sessions for people at GCHQ who were going through change. The sessions explained why our brains find change hard (Chapter 3 of this book), and also explored some of the small actions leaders can take to help get their teams 'back on track'. This section was based on some of the ideas in Chapter 4. People were also given small cards they could take away to remind them of the key practical things we can do to help people through times of uncertainty. The focus of these sessions was on asking participants what they would now go out and actually do.

## 4T days of positivity (4Ts)

The 4Ts – explained below – were inspired by several key concepts in this book, including:

1 People can think, work and collaborate better when they are in a 'reward' state and not in a 'threat' state.
2 Small actions can make a big difference.
3 Small steps feel less daunting to the brain.
4 Brains like novelty – not knowing quite what others will share provides interest.
5 Sharing ideas with others and acknowledging their contribution helps to create an 'ingroup'.
6 Habits take a while to form.

Katharine piloted the 4Ts with a group in human resources. As with so many good ideas, there's a beautiful simplicity to this concept and it's not onerous. The 4Ts stand for:

1 **Today's achievement.** This T asked a member of the team to share with others something that had gone well during the day. Choosing to remind ourselves about the positive things that we have done is rewarding to the

brain. Katharine also commented that sharing what one member of the HR team had achieved was important for knowledge-sharing across the HR team, as members can be working on quite disparate projects and would not necessarily pick up on what others are doing.

2  **Thank you.** This provided a nudge to a member of the team to think about who had helped them during the day and to say thank you for the assistance. The reason for a 'thank you' being sent ranged from someone saying 'hello' with a warm smile through to thanking someone for training them to use an online tool.

3  **Team tip.** This one encouraged people to share a useful nugget of information with others. Some team tips touched on well-being and health, for example: reminders to get up out of the chair and go for a walk around the building, through to pointing people to a really useful reference document of which people might be unaware.

4  **Tomorrow's nomination.** This last T passed the baton on, so that another team member would provide the 4Ts on the following day.

Playing with the name 4T, the team decided that 4T should run for 40 days, ie 8 weeks of 5 days. Eight weeks is a good number as the brain can form new habits in this time. After the initial pilot in HR, the concept was picked up by other teams. Katharine wrote a blog on the subject that became one of the most viewed blogs on internal media. People commented that the idea was 'brilliant', 'simple', 'not time-consuming', and, 'This has really cheered me up'. Katharine's advice to others thinking about implementing 4Ts is to think about the size of the group (not too large) and how much to 'push' the idea if some people are hesitant about it. She said everyone loved the idea by the end of the 40 days, but some were more reluctant to participate at the start. She also noted that the tip section was the one people found hardest, but some people widened the scope and shared tips about parenting, for example. People are busy and the other dilemma was how long to allow if the nominated person had not shared their 4Ts for the day. She said a gentle reminder to the person from her provided the nudge that they needed. In the parts of GCHQ that have used the 4Ts approach, it has become part of the vocabulary. When someone shares a useful piece of knowledge, people have been heard to say, 'That's good – it could have been a 4T'.

## *Helping leaders lead through change*

Charlotte commented that understanding more about the brain and what it needs during organizational change has also been usefully shared with others in the organization. Points that were particularly helpful for leaders to hear were about the brain's need for certainty and a sense of some control. Leaders were reminded that during change it's useful to be clear with people about what might be changing, but also what will stay the same, so that people know where they can continue as before. Also, on this point, GCHQ leaders kept in mind that it is important to treat the past with respect: people need to feel that their past ways of working are not being criticized or undermined, as this would likely generate a threat state.

Leaders learned that when we are in a state of uncertainty our brains can find it difficult to grasp new information: we are distracted and don't take in everything we are told. Leaders were also reminded that their reality is not *the* reality (remember those concave and convex circles in Chapter 2?) – all our brains filter information differently, depending on our past experience, expectations and our personal circumstances. It was suggested to them that they should keep putting themselves in the shoes of their team, so to speak, to try to look at the changes from others' point of view.

Also, bearing in mind how hard the human brain finds uncertainty, leaders were reminded that, on the whole, the human brain can deal better with bad news than no news.

## The language of science

Leaders at GCHQ tend to be intellectual and highly analytical. Concepts such as 'employee engagement' can be seen as 'pink and fluffy'. However, bringing science and evidence into sessions about people and performance proved appealing to these leaders (as it did with the banker mentioned at the beginning of Chapter 1). They were comfortable with, and interested in, a cerebral approach to talking about leading people.

Bringing scientific research and the language of neuroscience into sessions with leaders has proved to be a breakthrough. Most importantly, some people in GCHQ see a real positive difference in their leaders, to the extent that trust in leadership has increased by an extraordinary 32 per cent in parts of the organization where Lynne and the team have been applying this learning. Something is clearly working.

Other research revealed to leaders how much their behaviour began to set the tone and approach for others. In one part of the organization, the behaviour of the leadership team was perceived to be consistently 'task-focused'. Profiling revealed that 25 per cent of leaders were not predominantly task-focused but were behaving in this way to fit in with others.

## Social pain is real pain – ingroups and outgroups

As with so many groups, one of the points that had stayed with the GCHQ team was the fact that, to our brains, social pain is not so very different from physical pain – the same parts of the brain process both. From the brain's point of view, both matter in terms of survival.

The team shared this learning with leaders and it really struck home just how sensitive our brains are to social rejection and its impact on our ability to think (Chapter 5). It made them all the more aware of the importance of not having obvious favourites in the team as this would damage the productivity of those who felt themselves to be in the 'outgroup'.

## Practising mindfulness

Chapter 6 of this book explores how we can best manage our emotions during periods of change. As we know, one of the reasons we need to learn to regulate our emotions is because they are contagious. If we are feeling anxious or stressed, we are probably causing others to feel the same. As mentioned in Chapter 6, mindfulness is a technique that has been researched by scientists and continues to be the subject of studies. Practised over time, it has been shown to provide a wide range of benefits. A group now meets together at GCHQ to practise mindfulness, and it has become part of their approach to improving resilience.

## Communication and storytelling

GCHQ has also been applying some of the other ideas in Chapter 6 of this book. Leaders have been encouraged to tell personal stories while going through change, providing a more personal and empathetic approach to dealing with its demands.

Katharine told how she had applied both the learning in Chapter 6 and some of the insights from Chapter 5 – 'Our social brains' – to influence the way she communicated with a team going through transition. The work of Carr and Walton on the use of the word 'together' (Chapter 5) made a particular impression (Carr and Walton, 2014). In her role as transition manager, Katharine sent out a weekly message about the changes. First of all, she made sure the message went out without fail every week, as this provided people with certainty and predictability about the communication. She used the words 'our transition' and 'we' in her communications so that people felt they were in the transition together. She told personal stories about how the changes had affected her, knowing that personal stories and the emotions expressed in them can be more memorable than lists of facts about change. She was also very clear about what was known and not yet known about the transition so that the communication felt honest and transparent.

## Growth mindset

Having read the first edition of this book, the book club then moved on to reading Carol Dweck's *Mindset* (2007), referred to in Chapter 4. Dweck's work reinforces the point, just as neuroplasticity does, that we can continue to learn and improve throughout life. Many people at GCHQ have succeeded academically and are used to being seen as amongst the brightest. Dweck's book raised a pertinent question: is it all the harder for those people to fail at something? It also raised an interesting question around the job titles of those referred to as 'future leaders'. Might being put into this group also make people feel that they should not fail? Might the very title therefore inhibit them? This prompted the team to run a seminar about learning and the importance of learning from mistakes.

## Appreciative inquiry

Chapter 6 touches on appreciative inquiry (AI) – learning from what we have done well. Charlotte created workshops based on this approach, getting people to ask themselves, 'What are we good at here?'. This helped to shift an analytical and, at times, critical culture, to a more positive mindset. This positive shift was exemplified through the 'check-in' and 'check-out' process at the beginning and end of one workshop. In the check-in, one participant said he would rather be having root-canal treatment than be in the workshop. By the end of the AI session he said it was the best workshop of this type that he had ever been to and that he felt they had made real progress – a good example of how AI can be used to help shift people to a more positive mindset.

# The longer-term impact of learning about neuroscience

As mentioned at the start of this chapter, in 2014–15, leaders from four organizations participated in a one-day 'masterclass', then a follow-up call to answer questions and to discuss what the leader was going to put into practice. This was followed four months later with a half-day workshop in each organization to hear what participants had done and what differences they had seen. We measured both the learning and reactions of the leaders and of their direct reports, before and after the sessions. The workshops covered areas that will by now be familiar:

1 key points about our brains: neuroplasticity, our brains being prediction machines and the fact that our brains have not changed that much since we were on the savannah;

2 why our brains don't like organizational change;

3 performance: what helps the brain to be focused and creative, and the importance of getting the balance right between stretching people and not causing overwhelming stress;

4 the social brain;

5 staying calm under pressure;

6 planning and using the SPACES model;

7 hints and tips on getting the best out of the brain each day at work.

As a consequence of the workshop, the 40 or so leaders put all sorts of ideas into practice. Some leaders focused on improving their own performance, whereas some focused on other people, some a combination of the two. The ideas they acted on included ensuring that all team members and in particular new joiners felt they were a member of the 'ingroup', and in one case, relocating team meetings to ensure more face-to-face contact. Some changed their communication to ensure there was more clarity and therefore certainty about the purpose of it. One picked up on laughter being a great de-stressor for the brain and that emotion makes learning more memorable. He showed a branch a video of Bollywood dancers and made the point that if Bollywood can get 200 people into the right place at the right time, the branch ought to be able to place a few employees in the right place.

Some focused on improving their working day by blocking out 'thinking time' or by practising mindfulness to keep them calm and focused during the day. Others started to write their list of priorities first thing in the morning, before getting caught up in e-mails. One said this gave him the confidence to weigh up the importance and urgency of requests made of him during the day and to say 'no' or 'not yet'.

One of the most interesting areas was the difference the sessions had made to performance appraisal discussions. An HR business partner at Lloyds Banking Group said:

The single biggest takeaway was the learning around the continuum of reward to threat and the way the brain will react to those different approaches. For many years we have taught that performance management needs to be about confronting hard issues, giving tough feedback and raising performance by facing into performance issues. The (neuroscience) thinking turned my approach to performance management upside down. So, following the training I decided to change my approach, and focus on the positives. This didn't mean ignoring issues, but it meant accentuating the positives in our performance conversations. I was delighted with the results. I saw a big lift in the self-belief and confidence of two of my team, and as a result got more drive, energy and productivity from them both.

Two leaders who attended the masterclass were part of a taskforce to try to reduce costs. Before the masterclass the cost–performance index had been flat-lining for a long time. Following the masterclass, the two decided to change their approach. They thought about threat and reward and the inverted U of performance and how they could apply these to the taskforce. They became more open with information and talked more about a shared goal. They considered autonomy and made sure the group was clearer about what they could influence. They created an environment where people could openly share their thoughts. They set short-term goals and celebrated achievement. Over the following months, for the first time in years, they saw the cost–performance index begin to move. It improved by 70 per cent in the first two months:

> In what was a very difficult challenge in an environment that historically would have been highly tense and stressful, we now see a transitioned working group, working together and enjoying it while still delivering commitments to a strict timeline. … Focusing attention gives results … empathetic understanding and positive recognition of contribution has helped.

## Taking a longer view

For this edition of the book, I spoke again to three of the participants from those 2014–15 workshops. Let's hear first from Louise Wadman, formerly Head of Creative Services at Lloyds Banking Group:

> Applying neuroscience thinking has improved the communication strategies I deliver for leadership teams, especially during times of change. A few years ago, one of the banks I worked for wanted to re-orientate its culture. My communication team needed to ensure leaders quickly understood the

new cultural aspiration, supported it and could drive it forward. We did this through regular face-to-face activities.

The launch event went very well. After a short overview and update by the managing director (MD) followed by a question-and-answer session, we invited the 80-odd leaders to identify and agree on five signature actions. The idea was that the leaders could implement the actions in the branch network the following day with no additional resources. Via a series of facilitated sessions in small and larger groups, the team agreed the actions. It was an energetic and energizing event. From a neuroscience perspective, people may have arrived feeling unsure about things, but we enabled them to understand what was happening, reach their own insights, feel comfortable and get involved. The leaders left with positive 'towards/reward' mindsets. They were committed to implementing what was agreed and knew they would meet again soon. In the meantime, the change programme team, of which my team was part, spent the next couple of months working across numerous work streams with lots of deadlines and deliverables.

Eight weeks later, as soon as the first leaders started to arrive at the next event, it was clear that something was wrong. The atmosphere was tense and it did not improve during the keynote speech from the MD. At the first break, once I started to talk to the leaders, I identified the problem. The delegates thought they were coming to the event to be told there would be job losses. As soon as they realized this was not the case, the atmosphere lifted. The leaders relaxed and started to ask questions, became engaged and were receptive to the planned role alterations.

I learned so much that day. Although leaders had left the launch event with positive mindsets and in a 'reward state', as the weeks went by rumours started about planned changes to their roles. In the absence of any 'official' communication, and faced with uncertainty, they moved to a threat response and feared the worst. At the second conference they did not listen to the update because they were waiting for the news about job losses.

Between the two events, my team had neglected the leaders. Although the leaders themselves had been implementing the five signature actions from the launch event, we had not kept them up to date with what the programme team had been working on. Instead, the communication team was focused on what needed to be in place before the next conference. We had not understood that in order for a 'reward state' to be maintained, we needed to provide regular information to leaders and find ways for them to be involved. In other words, we had not understood that a 'reward state' needs to be nurtured, otherwise it will disappear.

Next let's hear from Tony Williams, who was working at Orbit Housing Group back in 2014:

> Since leaving Orbit and setting up my own consultancy, I have built on the neuroscience project. The key area has been executive mentoring and coaching. The knowledge of how neuroscience underpins the way we work through helping our brains perform at their best, coupled with using the SPACES approach (Self-esteem, Purpose, Autonomy, Certainty, Equity, Social connection – see Chapter 9) has made a big impact on the executives I mentor and coach. I would pick out the following highlights:
>
> - A better understanding by the executives and myself of their issues, drivers and concerns and how best to tackle these.
>
> - Allowing a different type of conversation during our sessions which is based on feelings and thinking rather than just a focus on what they have done in their work areas. This has also been supported by discussions around automatic negative thoughts (ANTs) and how to shift thinking by creating alternative thoughts (CATs).
>
> - Realization of the importance of social interaction in work and the effect it has on their thinking and performance.
>
> - The impact that executives can have when they fully and genuinely engage with their direct colleagues, teams and people across the organization.
>
> I would also emphasize the benefit of linking neuroscience thinking with psychometric profiling. This helps to provide more rounded pictures and knowledge of how an individual functions, why they act as they do and what strategies will help them going forward. The feedback from the executives I coach, their boards and direct reports has been very positive and this reaction has to a large extent been due to the influence of neuroscience (their words not mine!).

Now let's hear from Elizabeth O'Neill, who at the time of the research worked for the UK government department that is now known as Department for Business, Energy and Industrial Strategy (BEIS):

> Two things struck me from the applied neuroscience sessions: the first was the extent to which my judgement and performance are susceptible to variations depending on the context in which I am operating. We considered the best ways in which I can seek to bolster and protect myself from stressful circumstances to promote better balance and performance – for example, using mindfulness as a personal tool to assist in overcoming negative external factors. The second was the extent to which we all need to ensure that others in the team are not

excluded by our micro-behaviours, and the impact it may have on others if we do not manage these appropriately.

As for what I am doing differently as a result of the neuroscience sessions, two things stand out, one on a personal level, the other at a team level. First, on a personal level, using techniques such as mindfulness, but also ensuring that I am more self-aware in terms of physiological needs and responses to stress: for example, eating well and recognizing that prolonging working time could actually be counterproductive. Second, on a team level, showing increased consideration and awareness of the needs of others and the importance of inclusive behaviour.

Since the training, I have had better people engagement scores and tend to receive approaches from various colleagues as a confidante. This is definitely because I am, I think, viewed as more inclusive. Other outcomes as a result of learning about the brain: my personal well-being and resilience has been enhanced. I take work home less. I worry less. For my teams, it has, in my view, helped me build successful and empowered teams. My 360 feedback has been very positive.

The feedback from people – both those who participated in the masterclass years ago and the bankers, civil servants and technologists who have more recently learned about neuroscience – is overwhelmingly positive. For some it explains so much of what they have experienced on a personal level when going through organizational change. As many have said, there is something very reassuring about knowing that we are not alone in experiencing the social pain of not feeling part of the group or the distraction of uncertainty.

Neuroscience provides a lens through which to look at and understand people and what might be driving their behaviour. It also brings empathy – our brains are brilliant things but they have certain limitations. So much of what we experience in life and at work is about how our brains interpret and filter on our behalf. As I said earlier, we have more control over this than we probably realize. We can become more aware of how our brain interprets and what it needs to thrive and be productive; we have choice over the filters we set. Applied neuroscience is practical – small actions can make a big difference. As one leader summed it up, 'It's a game-changing concept'.

# The challenge ahead

Change theories and models abound. Neuroscience provides a science-based approach, which appeals because it provides a harder edge to the arguments about how to lead and support people during change. It is still in its infancy, but for the great majority of people who have participated in workshops, seminars and webinars, it has brought both insight and practical actions. One of the challenges is how to reach more people and how to get a critical mass of leaders who understand the brain a little better.

One of the benefits of learning about what helps the brain to perform at its best is that each person can go away and immediately put some of the learning into action. Leaders do not have to wait for the whole organization to move or for culture change. Small actions from each individual can make a big difference. Nevertheless, neuroscience raises questions not just about how each of us works with others, but also about the very way in which 21st-century organizations are set up and operate. From what we know, our brains flourish when the reward network is activated, when we feel challenged but not overwhelmed, when we have a perception of some control over our work, when we are not bored, when we feel we belong and when we are not always required to be 'on'. There is much for us to learn about our brains and the implications for how we work and thrive. These are exciting times. We have not designed organizations to get the best out of our brains – not yet, but we are learning.

## Summary of key points from this chapter

- Applied neuroscience is still in its infancy, but once people learn about it, it provides them with a lens through which to understand themselves and their co-workers better.

- In all cases, measurement demonstrated a positive shift for participants and, where we measured it, their teams could also see improvements.

- Some of the key features of these workshops and other interventions that enhanced learning included:

  - raising people's curiosity through initial communication about the learning;

  - asking people to put mobile devices away so that they could concentrate;

  - spacing learning over time so that it became embedded;

  - asking people to recall what we had covered in previous sessions to strengthen the learning;

  - enabling people to reach their own insights;

- giving people choice about what they acted on;
- having some fun and experiencing some of the concepts (emotions are important for making points memorable);
- encouraging people to take hand-written notes;
- getting people to think about what points they would share with others (helps the brain to retain information);
- using a range of techniques, both in the sessions and between them;
- helping people to understand what aids behaviour change (see research in Chapter 10: people who understand how habits are formed are better equipped to change them);
- encouraging people not to try to change too much at once;
- asking people to state during the workshop what they were going to do and suggesting they start to act on that goal immediately to create some momentum;
- follow-up reminder e-mails that were relevant to the goals people had set;
- measurement.

• Rooting sessions in science and evidence makes them persuasive to even the most sceptical of leaders.

• Years later people can still identify the difference learning about their brains has made to them.


# References and further reading

Carr, P B and Walton, G M (2014) Cues of working together fuel intrinsic motivation, *Journal of Experimental Social Psychology*, 53, pp 169–84

Dweck, C (2007) *Mindset: The new psychology of success*, Random House, New York

Kirkpatrick, D L and Kirkpatrick, J D (2009) *Evaluating training programs*, Berrett-Koehler, Oakland, CA

# INDEX

Note: Numbers, acronyms and 'Mc' are filed as spelt out. Page locators in *italics* denote information contained within a figure or table.

dog agility competitions (dog owners) 87–88
Donne, John 60
dopamine 11, 19, 29, 38, 40, 41–42, 48,
        100, 141, 192
    and goal-setting 47, 143, 194
    and mental fatigue 219
dorsal anterior cingulate cortex (dACC) 29,
        30, 62–63, 83, 144, 145
Dow Chemical 99
dreaming 226
Duhigg, Charles 192–93
Duke, K 224
Dunbar, Adrian 61
Dunbar, Robin 16
Dweck, Carol 43–45, 52, 99, 170, 171,
        175, 255

e-mails 19, 39, 49–50, 79, 157, 223,
        228–29, 244, 249
East model 203–04
Edison, Thomas 226
Education Endowment Foundation 5
*Education and Neuroscience Initiative* 5
ego 117, 131, 202, 203, 218, 219
    ego depletion 218, 219, 236
Eisenberger, Naomi 63, 67, 69, 71, 148,
        171, 173–74
electrical signals 10, 11, 16, 17, 18, 45
electroencephalography (EEG) 10, 11, 46
11 September 2001 attacks *see* Twin Towers
        (World Trade Center) attacks
Ellenbogen, J 226
emotional control 38, 82, 84, 90, 91–92
    *see also* self-control
emotional intelligence (EQ) 22, 60, 97
emotional reaction to change 188–89
emotional regulation 79–111, 130, 151, 184
emotions 32, 79–111, 114, 151, 158, 203,
        222–23, 229
    grabbing attention 149
empathy 69, 97, 99, 103, 132, 147, 153,
        254, 257, 260
employee engagement 28–29, 37–38, 68,
        71, 168, 169–70, 179
employee involvement 149–51, 155–56
employee performance 29, 37–57, 85,
        102–03
employee voice 169, 170
endings 172, 177
energy 21, 114, 116, 167, 188
    mental 89–90, 191, 224, 227, 228–32,
        237, 250
energy conservation 21, 116, 167, 188
*Engage for Success* 169, 179
Enterprise Transformation 248–50
epilepsy 7, 9

episodic memory 9, 83
EQ (emotional intelligence) 22, 60, 97
equity 175–76, 178, 259
Erasistratus 7
esteem 61
    *see also* self-esteem
ethics 41, 118, 145, 198
eureka moments 45–46
Evans, G 223–24
event-related potentials (ERPs) 45
existing routines 204
expectation of reward 42
expectations 33, 42, 142–43
expertise 118
Eysenck, M 221

face-to-face contact 69, 123, 132, 146–47,
        223, 244, 256, 258
    *see also* one-to-one meetings
failure 43, 45, 173
    *see also* mistakes
fairness 63–64, 71, 114, 127, 130, 158,
        170, 175–76
Falk, Emily 196–97
fast thinking *see* System 1 thinking
fear 7, 29, 82, 84, 122, 189
Febreze 193
feedback 50, 73, 196, 205, 208, 210, 211,
        242
Festinger, L 143, 199
fight or flight response 30–31, 38, 83, 84,
        113–14, 144
fighter pilots 23
filters 153, 158, 231–32, 253, 260
financial rewards 206
Fiske, Susan 120–21
five stages emotional regulation model 92–96
'five things' of mindfulness 100
fixed mindset 43–45, 48, 52–53, 208
flashbulb memories 149
flock and freeze response 30
flow 40–41, 171
focus 29, 32, 38–40, 46–47, 146, 190
four enablers of engagement 169–70, 179
4Ts of positivity 251–52
frazzled brains 39, 216, 222
Frederick, Shane 21, 116
frontal lobe 8, 24, 80
functional magnetic resonance imaging
        (fMRI) scanners 10, 12, 29, 63,
        127, 148, 196, 197, 225

Gage, Phineas 7–8, 80
Galen 7
Galinsky, A 121, 125, 139
Garner, R 144